T0320149

GUIDELINES FOR PROCESS SAFETY ACQUISITION EVALUATION AND POST MERGER INTEGRATION

This book is one in a series of process safety guideline and concept books published by the Center for Chemical Process Safety (CCPS). Please go to www.wiley.com/go/ccps for a full list of titles in this series.

GUIDELINES FOR PROCESS SAFETY ACQUISITION EVALUATION AND POST MERGER INTEGRATION

Center for Chemical Process Safety
New York, New York

A Joint Publication of the Center for Chemical Process Safety of the American Institute of Chemical Engineers and John Wiley & Sons, Inc.

Published by John Wiley & Sons, Inc., Hoboken, New Jersey.
Published simultaneously in Canada.

Library of Congress Cataloging-in-Publication Data:

Guidelines for process safety acquisition evaluation and post merger integration.
p. cm.
"A Joint Publication of the Center for Chemical Process Safety
of the American Institute of Chemical Engineers
and John Wiley & Sons, Inc."—
Includes index.
ISBN 978-0-470-25148-5 (cloth)
1. Chemical industry—Management. 2. Chemical industry—
Safety regulations. I. American Institute of Chemical
Engineers. Center for Chemical Process Safety.
HD9650.9.G85 2010
660.068'5—dc22 2010005157

10 9 8 7 6 5 4 3 2 1

CONTENTS

ONLINE FILES ACCOMPANYING THIS BOOK

This book is accompanied and complimented by two spreadsheets or tools you can load onto your computer and use to assist with building your own M&A process safety toolkit. These spreadsheets provide:

- Checklists of process safety issues that should be investigated or addressed in a proposed acquisition or merger,
- A draft of a possible integration plan of activities a reader or user of this Guideline may have to complete as part of merging together the process programs of newly acquired facilities with their own current operations, and
- A draft integration budgeting tool, to help with estimating the resources and costs that an organization may have to make provisions for when integrating two process safety programs together.

To access the spreadsheets, go to:
http://www.aiche.org/downloads/CCPS/Mergers_and_Acquisitions.zip

To unzip the files for use, enter the password: CCPS_Mergers

Email ccps@aiche.org with questions.

[illegible]

[illegible]

To access the spreadsheets, go to:
http://www.aiche.org/[illegible]

To access the files [illegible]

Email ccps@aiche.org with questions

Acronyms and Abbreviations

ACC	American Chemistry Council
AIChE	American Institute of Chemical Engineers
API	American Petroleum Institute
CAPEX	Capital Expenditure
CCPS	Center for Chemical Process Safety
CFR	Code of Federal Regulations
COMAH	Control of Major Accident Hazards (UK HSE Regulation
CSB	U.S. Chemical Safety and Hazard Investigation Board
DD	Due Diligence
EPA	U.S. Environmental Protection Agency
EU	European Union
HHC's	Highly Hazardous Chemicals
HSE	Health and Safety Executive (UK)
IPO	Initial Public Offering
IT	Information Technology Systems
JV	Joint Venture
KPI	Key Performance Indicator
MI	Mechanical Integrity
M&A	Merger and Acquisition
MOC	Management of Change
NFPA	National Fire Protection Association
NGO	Non-Governmental Organization
NMRIS	Near Miss Reporting Information System
OPEX	Operating Expenditure
OSHA	U.S. Occupational Safety and Health Administration
P&ID	Piping and Instrumentation Diagram
PDF	Portable Document Format
PHA	Process Hazard Analysis
PS	Process Safety

PSIC	Process Safety Incidents Count
PSISR	Process Safety Incident Severity Rate
PSTIR	Process Safety Total Incident Rate
PSM	Process Safety Management (U.S. OSHA Regulation)
Q&A	Question and Answer Process
RAGAGEP	Recognized and Generally Accepted Good Engineering Practice
RBPS	Risk-Based Process Safety
TCPA	Toxic Catastrophe and Prevention Act
TQ	Threshold Quantity

Glossary

Acquisition	Is the purchase of assets or equity interest in a company and may in either case include the transfer of the operating staff from a seller to a buyer.
Decapitalization	Is the sale of the assets of a business, where the brand or operation of the facility remains with the seller.
Divestment	Is the sale of assets or equity interest in a business entity and may include the transfer of the operating staff from the seller to a buyer
Due Diligence (DD)	Is a term used for a number of concepts describing the performance of an investigation or assessment of a business or of certain assets and may address one or all of accounting, legal, engineering, environmental, and other analyses to evaluate the assets, liabilities, and potential liabilities of a business. The term commonly applies to voluntary investigations
Hazard	A chemical or physical condition that has the potential for causing damage to people, property or the environment. A hazard is intrinsic to the material or to its conditions of storage or use. With respect to chemicals, "hazard" may include toxicity (acute or chronic), flammability, corrosivity or reactivity.

Initial Public Offering (IPO)	An initial public offering is when a parent company packages one or multiple assets into a new company whose stock will be offered for sale to the public.
Integration	For the purpose of this Guideline, integration is the process of planning and implementing various activities to merge together the operations, systems and staff of the newly acquired assets in accordance with the desires and expectations as established by executive management
Joint Venture (JV)	A joint venture (JV) is the combining of assets and/or cash into a new company that is an independent legal entity from the companies forming the JV
Materiality& Material	In the context of an M&A is the measure of the significance or effect that the presence or absence of an item or issue may have on the transaction.
Merger	Is technically the combination of two business entities where typically one entity survives and the other comes to an end, but with the practical result that the assets, staffs and managements (some or all) become combined into a single, larger company.
Process Safety (PS)	A disciplined framework for managing the integrity of operating systems and processes handling hazardous substances by applying good design principles, engineering, and operating practices. It deals with the prevention and control of incidents that have the potential to release hazardous materials or energy. Such incidents can cause toxic effects, fire, or explosion and could ultimately result in serious injuries, property damage, lost production, and environmental impact.

Process Safety Culture The combination of group values and behaviors that determines the manner in which process safety is managed. A sound process safety culture refers to attitudes and behaviors that support the goal of safer process operations.

Process SafetyManagement The application of management systems to the identification, understanding, and control of process hazards to prevent process-related injuries and incidents; it is focused on prevention of, preparedness for, mitigation of, response to, and restoration from catastrophic releases of chemicals or energy from a process associated with a facility.

Process Safety Metric A standard of measurement or indicator of process safety management efficiency or performance.

Risk A measure of potential loss (for example, human injury, environmental insult, economic penalty) in terms of the magnitude of the loss and the likelihood that the loss will occur.

Stakeholder Individuals or organizations that can (or believe they can) be affected by the facility's operations, or that are involved with assisting or monitoring facility operations.

Swaps Asset swaps are the trading of similar assets of similar value from one company to another, usually because of geographic and/or business synergies.

Toll Processor Agreements Are contractual agreements between a company and supplier of manufacturing services, branded products or proprietary feedstocks.

Vendor Due Diligence (VDD) A VDD is typically conducted by the seller using a third party consultant. The VDD provides potential buyers with a brief history and current summary of the facility's health, safety, environmental and process safety programs, compliance with applicable regulations and known liabilities from the Seller's perspective.

Acknowledgments

CCPS wishes to acknowledge the many contributions of the ioMosaic members who wrote this book especially the principal author and editor Dr. Gary Kenney and authors Messrs. Henry Ozog and Mr. George Groves. The authors wish to thank the following ioMosaic personnel for their technical contributions and review: Dr. Georges Melhem, Ms. Susan Ozog and Ms. Vanessa Millette.

The American Institute of Chemical Engineers (AIChE) and the Center for Chemical Process Safety (CCPS) express their appreciation and gratitude to all members of the Process Safety M&A Subcommittee and their CCPS member companies for their generous support and technical contributions in the preparation of these *Guidelines*.

PROCESS SAFETY M&A SUBCOMMITTEE:

Kathy Anderson, Chair	Vertellus Specialties, Inc.
Steve Arendt	ABS Consulting Group
Eric Freiburger	Nova Chemical
Bob Genau	DuPont
Rick Griffin	ChevronPhillips Chemical Co.
John Herber	3M
Jim Muoio	LyondellBassell
Bob Ormsby	CCPS
Karen Person	CCPS
Glen Peters	Air Products
Jatin Shah	BakerRisk
Kenan Stevick	Dow
Karen Tancredi	DuPont
Bob Perry	CCPS Staff Consultant

Before publication, all CCPS books are subjected to a thorough peer review process. CCPS gratefully acknowledges the thoughtful comments and suggestions of the peer reviewers. Their work enhanced the accuracy and clarity of these guidelines.

Peer Reviewers:

John Alderman	RRS/Schirmer Engineering
Graham Bennett	DNV Consulting
Robert Fischer	TOTAL
David Guss	Nexen
Dave Krabacher	Cognis
Steve Metzler	Primatech
Vince Murchison	Sonnenschien, LLP
Rich Novak	Cognis
Cathy Pincus	ExxonMobil
Adrian Sepeda	CCPS Emeritus

Preface

"Various studies have shown that mergers have failure rates of more than 50 percent. One recent study found 83 percent of all mergers fail to create value and half actually destroy value. This is an abysmal record. What is particularly amazing is that in polling the boards of the companies involved in those same mergers, over 80 percent of the board members thought their acquisitions had created value."□

Robert W. Holthausen, The Nomura Securities Company Professor, Wharton School of Executive Education

Courtney finally settled into her seat on the plane and closed her eyes. The past seven hours had been a mad rush.

What were they thinking about on the 26th floor of headquarters in Houston? Another acquisition? We haven't even gotten our heads around the acquisition of Independent Refining we bought three years ago! AND this time it's a chemical company! Bland Petroleum has no experience with bulk chemical processing.

Courtney was in the second day of an HSE audit of one of the former Independent refineries when she received a call from Houston to get back as soon as possible. The CEO of Bland Petroleum had just announced they were entering into an agreement to purchase White Hot Chemicals. A due diligence team was being formed and Courtney was to lead that part of the review into all HSE regulations and issues. The Team was meeting tomorrow morning and Courtney needed to be in Houston in time for the kick-off meeting.

When Courtney heard the news, her first thoughts were – "White Hot Chemicals, never heard of them". While in the lounge at the airport she was able to connect to the Internet and did a quick search on White Hot. She found they had six plants; two in Texas, one in California, one in New Jersey, one in Pennsylvania and one near Terneuzen, Netherlands. She also discovered one of the Texas plants had recently been audited by OSHA. OSHA found numerous process safety related violations and White Hot agreed to correct the various deficiencies in a 20-month period. That was eight months ago.

She felt the plane being pushed back from the gate and heard the engines start. Courtney turned her head slightly and looked out the window. It wasn't until she had been in the cab to the airport that Courtney had the chance to wonder – why me? Despite having been in the industry for eighteen years, I've never been directly involved in a merger or acquisition. Yes, her previous company had been bought but at that time she was relatively junior in the organization and it all just happened around her. Besides not much seemed to have really changed other than a number of their IT systems were replaced. As her mind returned to the present situation, she started asking herself: "What are they expecting of me? What are the first steps? What kind of questions or information should I be seeking? Where do I get it from? Is there any guidance that will help me through the next several weeks or, more likely, months?" Courtney began to feel exposed, lonely and yes a bit frightened at what lay ahead. One thing she was certain of – "I'm not going to sleep well at all tonight!"

Sound familiar? Been in that situation yourself? Statistics compiled by PriceWaterhouseCoopers (PWC) on merger and acquisition activity in the chemical industry indicate you have a lot of company. Over the ten-year period of 1998 – 2008, PWC reported the following numbers and value of acquisitions just within what is designated the 'chemical industry':

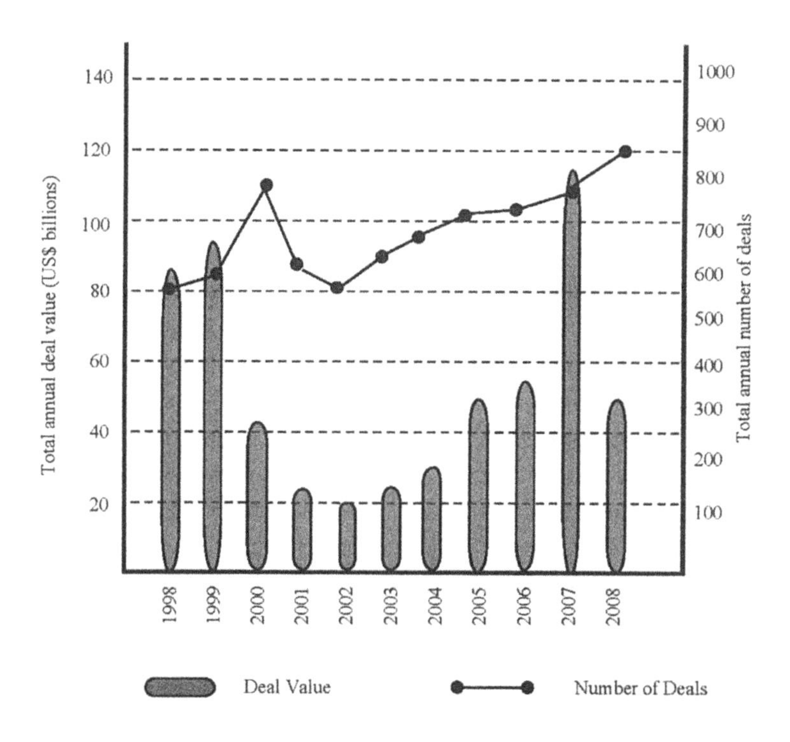

Figure 1. Chart of Reported Merger and Acquisition Activity in the Global Chemical Industry compiled by PWC

Because of all this activity, the Center for Chemical Process Safety decided to develop a guideline on process safety management issues that should be considered when an organization is contemplating a possible acquisition or merger. The purpose of this Guideline is to help individuals charged with responsibility for addressing process safety related matters to answer questions, concerns and issues that arise during the acquisition of or merger with a facility or company that has process safety related operations. In other words, questions as started to course through Courtney's mind.

This Guideline, of course, will not answer all the questions that will arise as an organization or an individual progresses through the various stages of a merger or acquisition. It is hoped though that it will assist the reader to better understand the various phases of the merger and acquisition process, the types of process safety issues that should be examined as the acquisition moves forward and finally lay a foundation for establishing an integration plan to bring what at times might be two disparate approaches to process safety to a single whole.

The Guideline has been constructed more to serve as a working tool than a textbook on the merger and acquisition process. As is particularly true in our profession in today's business climate - Who has the spare time or inclination to sit down and read a book on process safety related issues during a merger or acquisition unless that situation is staring them in the face? So if you are like Courtney and are facing an acquisition or possible merger this Guideline is targeted at you. It is our goal by taking the time to go through it, we will have reduced your anxieties, increased your confidence, helped identify the issues you need to explore as well as when to explore them and reduced the burden on you and your company's M&A team. And finally, we hope it will help guard against your merger or acquisition falling into that eighty percent that does not create value – at least from a process safety standpoint!

EXECUTIVE SUMMARY

WHY THIS GUIDELINE?

This Guideline is born out of the hard experiences of various members of the Center for Chemical Process Safety (CCPS). One of those hard experiences is that CCPS member companies discovered after purchasing a site or merging businesses a variety of process safety concerns or issues had to be corrected. In correcting those issues they found it added another ***ten to thirty percent*** to the initial purchase price. The adage that 90% of an iceberg is below the water line seemed to well illustrate what these companies experienced.

Why did this occur? It was a direct result of process safety not being an integral part of their M&A due diligence activities. It was not until after the deal closed and the integration process started that they were able to identify and then quantify the cost to correct these process safety issues. Only after they were able to develop hard data and present it to their management did these companies gain support to include assessing process safety to their due diligence activities. Support to develop tools and mechanisms that helped ensure process safety issues were identified and assessed early in the due diligence phase of a potential acquisition or merger. This Guideline was developed, then to:

- Pass that experience along to others to help them make that same case to their management (namely that process safety issues need to identified and investigated thoroughly in the due diligence process),
- Provide assistance on the various types of process safety issues that should be investigated, and
- Provide assistance with planning the integration of multiple process safety programs into your business after the deal has closed, and
- Help to use risk based experiences and expertise in the merger and acquisition process.

The Guideline with appendices is over 300 pages long. Recognizing you might not have the time to sit down and read through it all, this Executive Summary was prepared outlining many of the important points in each Chapter. Further, the Appendices contain three spreadsheets or tools you can load onto your computer and use to assist with building your own M&A process safety toolkit. The appendices provide:

- Checklists of issues to be addressed in a proposed acquisition or merger,
- A draft integration plan of activities that you may have to complete as part of merging the newly acquired facilities with current operations, and
- A draft integration budgeting tool, to help with estimating the resources and costs associated with the integration process.

These spreadsheets are available on the site http://www.aiche.org/downloads/CCPS/Mergers_and_Acquisitions.zip
Password: CCPS_Mergers

CHAPTER 1 – AN OVERVIEW OF PROCESS SAFETY

The first Chapter provides readers who are new to Process Safety (PS) or Major Hazard issues with some fundamentals in these areas. It also presents a few case studies of recent investigation(s) into major PS related accidents. These investigations examined changes to the organizational structures, management systems and the general overall approach taken toward process safety issues. One case study demonstrates that even where the acquisition involves two companies who have a history of operating similar if not identical plants or processes, their individual approaches to process safety may be so different, it is difficult to integrate seamlessly the two programs together.

For a definition of process safety, the current working definition as compiled by the CCPS is used:

> *Process Safety – is a disciplined framework for managing the integrity of operating systems and processes handling hazardous substances by applying good design principles, engineering and operating practices. It deals with the prevention and control of incidents that have the potential to release hazardous materials or energy. Such incidents can cause toxic effects, fire or explosion and could ultimately result in serious injuries, property damage, lost production and environmental impact.*

The Chapter points out that the effectiveness of process safety programs must be evaluated separate from occupational or personal safety programs. Trevor Kletz notes;

> *"Personal safety metrics are important to track low-consequence, high-probability incidents, but are not a good indicator of process safety performance. The lost time rate is not a measure of process safety..."*

You should be cautious then of using personal safety indices (e.g. lost-time-injury rates) as a determinant of the state of a candidate's process safety program(s). Leading and lagging performance measures have been developed to track and assess the state of process safety programs. In addition to any personal safety indices, process safety related indices should be requested and used for assessing the health of a target candidate's management of major hazard or process safety issues.

Process safety related accidents are sometimes referred to as being of low likelihood (i.e. occurring infrequently) but often having major if not catastrophic consequences. As the consequences of an accident becomes more severe, society and as a result, respective governments, find PS accidents to be unacceptable or intolerable. Control of such accidents includes effective leadership, administrative controls (e.g. management systems, training, procedures, etc.) as well as assuring the physical plant and equipment used to process or handle highly hazardous materials or processes are fit for their intended purpose. Effective process safety programs entail assuring the health of all such risk control measures.

The first step in assessing an acquisition candidate's exposure to process safety issues, is to identify the actual type and quantities of those materials stored or processed at a particular site. Lists of 'Highly Hazardous Chemicals' have been developed using a variety of physical and chemical characteristics (e.g. their flammability, toxicity, reactivity, etc.). Many of these lists have been adopted into regulations which set limits, above which a site or process are required to put into place process safety or major hazard accident programs.

In addition to facilities that handle highly hazardous chemicals, many processes or operations are prone to generating flammable or explosive dusts. These include plants or facilities handling or processing even common materials such as sugar, wood, grain(s), coal and many metal dusts such as aluminum. These too should be examined for process safety issues.

There are numerous process safety resources available to an individual needing further research, support or assistance in this area. These include:

- Government agencies – e.g. US-OSHA or UK-HSE
- Industry associations – e.g. the Chlorine and the American Petroleum Institutes
- Professional associations such as the American Institute of Chemical Engineer's – CCPS website, the Canadian Society of Chemical Engineers and the Institute of Chemical Engineers in the UK.
- Academic institutions such as the Mary K O'Connor center for Process Safety

A more exhaustive list of such agencies or groups and their websites is listed in Table 1 of Chapter 1 (page 39).

CHAPTER 2 –THE MERGER AND ACQUISITION PROCESS

Chapter 2 was written to provide a reader a basic understanding or overview of the merger and acquisition transaction process whether it involves the acquisition of assets or the purchase of company stock or the merger of two companies.

It is important to remember that an acquisition or merger always involves two parties, a seller and a buyer. A transaction typically entails an acquisition of either assets (such as facilities and process units) or equity in a target company by a buyer (equity can include stock, partnership interests and other forms), while on the other side of the table is a seller who is divesting itself of an asset or the equity in a subsidiary or affiliated company. Definitions for acquisition, divestment and a merger are provided as follows:

- ***Acquisition*** – is the purchase of assets or equity interest in a company and may in either case include the transfer of the operating staff from a seller to a buyer.

- ***Divestment*** – is the sale of assets or equity interest in a business entity and may include the transfer of the operating staff from the seller to a buyer.
- ***Merger*** – is technically the combination of two business entities where typically one entity survives and the other comes to an end, but with the practical result that the assets, staffs and managements (some or all) become combined into a single, larger company.

A brief overview of other types of merger and acquisition activities such as the forming of joint ventures, mergers, etc. are also discussed in Chapter 2.

These guidelines use the terms "merger and acquisition" and "M&A" process as a shorthand reference to a plethora of transactional structures only lawyers can conceive. But in the end, the net result is that processing plants, individual process units, manufacturing plants, transportation facilities, and other complex facilities change hands to a new owner and/or operator. Whether via an asset or equity transaction, the process typically begins with signing confidentiality agreements followed by a series of exchanges of information from the seller to the buyer concerning the subject. If business objectives persist and negotiations progress, due diligence typically occurs prior to and during drafting of agreements to govern the exchange. This process may occur over a period of time from as short as a few months to well over a year. However, while simple in overall form, the process can be highly resource intensive as well as physically and emotionally demanding.

A critical part of the M&A process is the due diligence phase. For the purposes of this Guideline due diligence is defined as:

> *Due Diligence (DD)* – is a term used for a number of concepts describing the performance of an investigation or assessment of a business or of certain assets and may address one or all of accounting, legal, engineering, environmental, and other analyses to evaluate the assets, liabilities, and potential liabilities

of a business. The term commonly applies to *voluntary investigations*.

Of importance is that both the seller and the buyer should conduct a due diligence examination of the assets in question. The purpose is to identify all relevant and material issues or liabilities, their estimated value and any conditions that could influence the final decision to acquire those assets. When performing such assessments it is usual to assemble a multi-disciplinary team. The role of the DD team is to:

- Raise potential issues to the M&A business team that could require heightened management attention,
- Provide the M&A financial team with a valuation of potential process safety costs, and
- Provide the M&A legal team with sale agreement language, mechanisms and concepts.

When performing a due diligence, it is vital to set a level of financial materiality that the DD team is to evaluate and for inclusion in their report. A general rule of thumb is to set that threshold at ten percent of the estimated purchase price for the group of assets being examined.

A checklist of potential process safety or major hazards issues that a DD team should consider is provided in the Appendices as well as on the accompanying files (see page xi for details on how to access these files.)

CHAPTER 3 – SCREENING POTENTIAL CANDIDATES

Whether you are called in at the earliest stages of a potential acquisition or later when a DD team is formed, a search on the Internet may provide you valuable background information on the target candidate(s). Start with the target candidate(s) company website(s). Next check websites of local, state or national regulatory websites to see if they contain any information on the particular site or company being considered. News articles from local, state or national news services or papers are also a useful source. Finally, you may want to check various Non-Governmental Organization (NGO) websites to see if they contain articles or information pertaining to the site(s) or company being considered. The nature of all this information should be noted and investigated further when you move into the later stages of the due diligence process.

From the company's websites, see if you can obtain information on their overall policies and approach to HSE in general, and process safety, in particular. Ideally you would want to obtain details on the company's approach to process safety or asset-mechanical integrity. Once in the company's website, do searches for such terms to see if it provides any references or 'hits'. Accident statistics or information on actual incidents should be searched for as well. The search should be balanced though, so see if the company has received any awards or commendations for their safety and environmental performance or practices.

A search of local, state and regulatory agency websites may turn-up information of recent enforcement actions, notices of violations, judgments or administrative agreements recently reached with the target company or site. Such matters should be brought to the attention of the M&A project lead if identified. Also search for the status of any permits the target company holds. The nature of such permits could prove material at this stage of the DD process.

Negative news articles or negative discussions of the company or site on NGO websites should be noted for future investigation. However, where it is felt the issue or issues underlying the reason for the news articles are of potential materiality to the proposed acquisition or merger, these should be brought to the attention of the M&A project lead as soon as possible.

Websites that may provide information as described above are presented in Chapter 3 (Tables 1, 2, 3 and 4) and in Table 1 of Chapter 1.

Where it is possible to obtain the location(s) of the actual sites that are to be acquired, these should be searched using Google Earth® or Microsoft Virtual Earth®. These websites provide the ability to view the site in relation to its neighbors, watercourses, railroads and adjacent facilities, i.e the surrounding general environment. Zooming-in on the site will allow you to gain insight regarding the number and general size of storage tanks, their location with respect to surrounding or neighboring facilities as well as whether there are sites of special interest nearby (e.g. schools, hospitals, other processing plants, etc.). Capturing such information will give the PS reviewer a head start on what issues they might need to examine as part of the site visits, or information they might request to be placed in the data room.

CHAPTER 4 – THE DUE DILIGENCE PHASE

The activities or steps involved with divesting a group of assets versus acquiring them are very similar. However, the order in which the activities will be carried out differs.

In either case the due diligence starts with developing or preparing a checklist of potentially relevant issues or concerns associated with the assets being considered for sale or purchase. As noted earlier, a checklist of potential process safety related issues that might be relevant to your particular situation is included in the Appendix to this Guideline. It is provided to assist you with developing your own unique checklist for the assets being considered for sale or purchase. Further the checklist you develop should be a 'living' checklist where items are added to or deleted from the checklist as the merger and acquisition process moves forward. There is little doubt that as the M&A progresses information will be identified and evaluated that discounts the need to investigate certain items while in other cases it might demand concerns be added to the checklist.

In a divestment, the seller will compile various factual information material to the assets from a variety of sources including their own intranet as well as through a set of site visits. The seller will then collate this information and:

- Use it to prepare an Information Memorandum which will be sent to various potential buyers or bidders,
- Populate a data room with the information, and
- Develop a more extensive document called a Vendor Due Diligence (VDD) report to provide the potential buyers (bidders) with key information about the asset(s) being offered for sale.

Information should not be created specifically for any of the above purposes. Where information is lacking, that should be addressed factually. Similarly when developing the information memorandum or the VDD Report these are to be written in as factual a manner as possible. It is important to highlight all related HSE issues or liabilities to the Seller's business team along with estimating their potential impact on the worth of the assets being considered for sale. The Seller's business team needs this information to assess fairly the various bids they will be receiving from bidders as well as any conditions a particular buyer attaches to the sale.

Data rooms today can take the form of a physical location populated with hard copies of materials or as a secure web or File Transfer Protocol (FTP) site. If you are responsible for providing material to a data room, develop a standardized format for the files that includes a means to assess the contents quickly. While there are a lot of frustrating things in life, one of them is to open an electronic folder of information only to be staring at hundreds of files with nothing more descriptive than something like SellCo.001.0001, SellCo.002.0015, etc. requiring you to open each file to determine whether or not it is relevant to your needs.

As a Seller, you may want to consider undertaking a 'reverse' due diligence exercise of the potential buyers or bidders. Depending on the level of confidentiality surrounding a particular divestment, this might have to wait until after a short-list of candidates is developed. This 'reverse' due diligence would take the form of an Internet search of the potential bidders as outlined in Chapter 3. Here as the Seller you are searching for information to determine the extent to which the bidders have experience operating sites subject to process safety or major hazard concerns. Where you find possible issues such as a bidder has not had past experience operating sites with process safety related concerns, they should be conveyed to the Seller's M&A team so they can factor those matters into their bid evaluations.

Buyers or bidders will also want to start with a checklist of potential issues and the results of the various Internet searches undertaken. These will form the basis of their requests or searches for information to be made available in the data room as well as issues to investigate if and when you undertake visits to the actual sites. Checklists will also help with maintaining a level of consistency where the assets being considered consist of multiple sites. This may mean mobilizing more than one team to visit the various sites owned by the seller.

The site visits are often viewed as being the most critical of all the DD activities undertaken. Pre-planning the site visits is vital, because normally the allocated time for the visits will be tight. Ideally for larger sites, the buyer(s) should try to negotiate a period of three or so days to complete their reviews per each site. On very large sites this may extend to five days while for smaller sites the review may be able to be completed in a matter of hours or a single day. However, for a variety of reasons the time per site may well be reduced to as little as a single day at each site even for the larger sites. Planning the visits is critical, to allow the team as much time at the actual site as possible, as well as talking to the actual staff at that site.

The ideal site visit will provide the due diligence team the opportunity to:

- Review relevant process safety related materials and documents, (e.g. engineering drawings, maintenance and inspection records of critical equipment, risk assessments, operating procedures, etc.)
- Conduct a walk-through of the site to visually assess the state of major equipment items (e.g. tanks, pressure vessels, pumps, compressors, relief and blow-down systems, etc.),
- Place the site in context with any sensitive sites or areas such as schools, hospitals and environmental receptors,
- Meet, discuss or interview not only key staff but a mixture of operators and maintenance staff as well.

To help with the eventual development of the acquisition Due Diligence report, a set of forms should be generated so that notes of particular issues can be captured quickly while on site. Then later in the day or in the evening those notes can be expanded-on while the information is still fresh in the teams' mind.

As a prospective Buyer your company will submit to the Seller's business team a monetary bid as well as general contractual terms and any conditions or limitations the Buyer(s) feel are necessary. As a general rule, your DD report should contain the following information as input to the final bid submission:

- Summarize the assessment process
    - Outline the activities completed,
    - Discuss the activities not undertaken or completed,
    - Discuss the potential implications of the items not completed.
- Highlight all material findings
    - Identify and summarize data and information
    - Define and to the extent possible, quantify gaps (i.e. differences between 'the way we do it' and 'the way they do it')
    - Identify and quantify regulatory compliance issues

- Discuss recommendations

  - Discuss the PS liabilities the target company, operations or sites may have and provide a range of estimates in monetary terms to address these issues.
    - Outline the resources/costs/budgets to address issues identified to this point in the process.
  - Describe why there is a need for further assessment and evaluation of the target company's approach to process safety.
- Conclude with a discussion of whether this is a good deal or not from a PS perspective

CHAPTER 5 – DEVELOPING THE INTEGRATION PLAN

Following the close of the deal, the second major phase of the M&A process begins – that of integrating the assets into the newly formed company, division or business unit of the new parent company.

A study conducted by Mercer Consulting found poor post deal integration was the primary reason that in forty-three percent of over 300 mergers reviewed, the newly merged businesses did *not* out perform their competitors. The need to plan the integration process well is obvious.

An overview of a suggested process for planning the integration of the Process Safety programs of the newly acquired assets or business with the new parent company's activities is presented in the diagram.

It starts with establishing an understanding and mutual agreement with senior and executive management on the boundaries that will surround the future operation of the newly acquired and combined businesses (i.e., the overall integration strategy). From

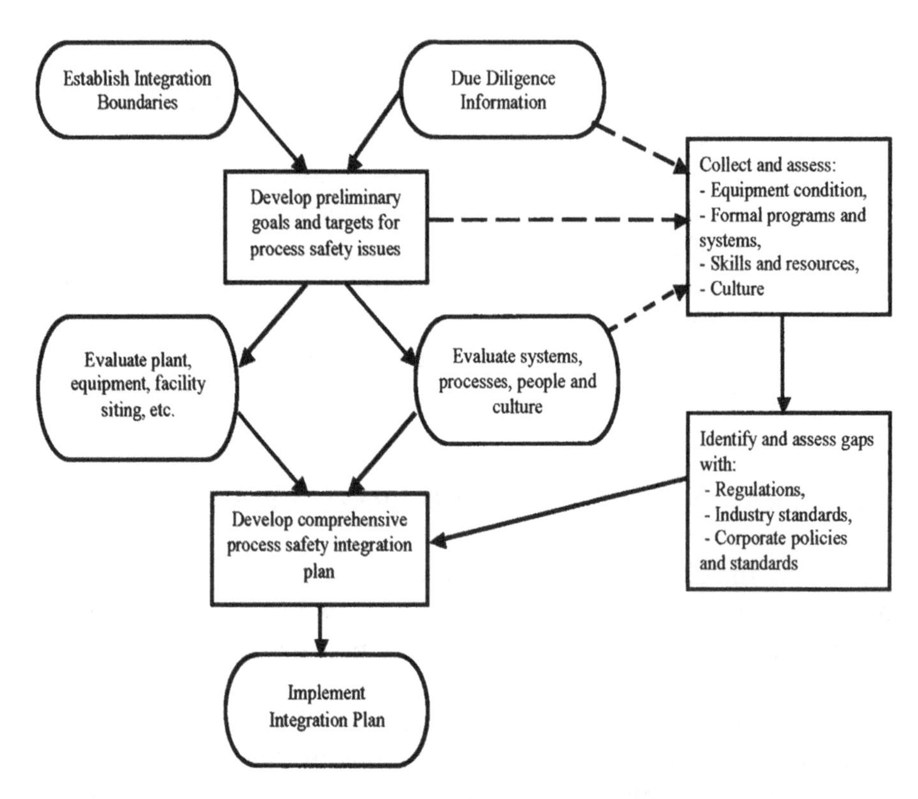

Figure E-1 Overview of the Integration Process

there it continues on through a process of review(s) to identify gaps and the extent of those gaps and culminates in the development of a resourced plan for addressing or closing all such gaps.

The first step in the integration process is to agree with senior management what boundaries they have set for the overall integration process and future operation of the newly acquired assets. In summary these might include:

- The 'extent' to which the newly acquired business or facilities are to be integrated (e.g. will they be operated as a stand-alone business or are they to be fully integrated?)
- The 'height of the bar' senior management is setting for the operation of the new business or facility with respect to process safety (e.g. is it compliance with applicable

regulations only or are they to be brought to the standards of the parent company), and

- The resources and budget that will be committed to achieve integration (i.e. will the integration process be granted an actual and credible budget or will it be a case of having to 'make-do'?).

The next step is to develop a set of measurable expectations in each of the four main elements that comprise a typical integration process. Expectations in each of the following areas should be clearly established and thereby provide a means to measure any gaps that may exist between the current state of each element and the desired state. This will also provide the ability to track on a more objective basis, progress towards achieving the expectations of management. The four elements include:

- The physical plant, equipment and operating processes,
- The human resources required to drive and support an effective process safety program,
- The desired state of the various management systems, programs and procedures that constitute an effective approach to process safety,
- The culture required to ensure these requirements are effectively carried through in day-to-day operations.

The third step of the integration process is to identify both the skills necessary and the amount of human resources that will have to be committed to the integration team(s).

There are likely to be two major phases to the integration process. A first evaluative phase that involves auditing or measuring the:

- State of the current physical plant and equipment,
- Management systems,
- The current performance of the site measured against a set of leading and lagging process safety indicators,

- Human resources involved with driving the process safety programs, and
- Culture towards process safety that exists within the management and supervisory levels as well as at the operator and craft level.

The second phase entails taking the corrective actions necessary to close any identified gaps. The technical skills required in the integration teams are likely to be fairly common across both phases, for example asset or mechanical integrity, risk assessment, human factors, etc. However in the first auditing phase these technical skills need to be supplemented with experience in the techniques of auditing such issues. In the second phase, rather than auditing skills the team members will need to be experienced in building and implementing corrective action programs. When establishing these teams, recognize the physical, mental and emotional demands on the integration team or teams will be considerable. The integration plan should, therefore, make provision to rotate team members where and when necessary.

The next activity will be to establish a baseline of actual process safety practice within the newly acquired business' various sites or facilities. Reviews of documents, programs and site visits, will need to be undertaken to gather enough information to assess:

- The general condition or mechanical integrity of the physical plant and equipment with a focus on the degree of compliance with any and all regulatory requirements or to a good industry standard,
- Whether the skills and technical expertise are currently in place to manage a process or major hazard safety program and the degree of internal versus external support that is used to support and maintain such programs,
- The state of the various management systems that are in place and form the framework for the newly acquired business or sites process or major hazard safety program, and
- The culture of the management, supervisory levels and operating and maintenance staff towards actual day-to-day

adherence to or implementation of an effective process safety program.

A protocol of process safety issues using input from the integration boundaries, the expectations, information from the due diligence teams, etc. will need to be developed. Review teams will need to be mobilized, arrangements made to conduct site visits, etc.

Following completion of the site visits, a baseline report will be developed that provides a clear picture of the gaps between the current state of the four main elements outlined above and the desired expectations for the future process safety program. The report should prioritize the gaps using a risk metric and critical gaps should be highlighted. First order estimates of the resources required to close the gaps should be provided or where this is not possible, what steps would be necessary to develop such first order estimates.

In consultation with the senior management team responsible for the overall integration or merging of the newly acquired assets, development of an agreed consolidated action plan is the final step in the integration planning process. This consolidated action plan will contain the following major parts:

- A program to address and put into place temporary controls within a period of ninety to one-hundred days for all risks deemed 'intolerable',
- A capital works program to address deficiencies or gaps found in the physical plant, equipment and processes.
- A plan to rationalize resources and organizational responsibility for process safety related activities,
- A plan to address required PS related training and competency requirements at various levels in the organization,
- A plan to address gaps in the policies, procedures, and systems that govern the desired approach that senior management expects the new organization to achieve.

- A Management of Organizational Change plan that assesses whether the planned changes will continue to control risks to acceptable levels and identifies the controls necessary to assure such risks continue to be controlled through the transition.
- A consolidated plan to manage the overall integration process, track progress towards goals, and implement corrective measures as soon as variances are identified.

CHAPTER 6 – IMPLEMENTING THE INTEGRATION PLAN

The consolidated action plan and its various elements as described in Chapter 5 will have laid out what you intend to change. However, how to go about instituting those changes is just as important.

Integrating or merging the programs, systems and activities that a newly acquired set of assets developed to manage process safety or major hazard issues with those of a new 'parent' company involves change. And in the eyes of the staff of the newly acquired assets, most probably, they will view that change as 'massive' and are likely to resist. They may ask – "why change?" A study of change programs instituted in large organizations by the management consulting firm KPMG found that 80% of all such change efforts failed to meet the expectations originally established by senior management. Professor John Kotter (Harvard Business School) and Dr Dan Cohen set out to study why such programs failed and found:

"More than any other single finding, we discovered in this second project that people changed less because of facts or data that shifted their thinking than because compelling experiences changed their feelings. ... Too many people were working on the mind without paying sufficient attention to the heart."

While the integration process may involve modifying equipment or changes to the formal written management systems, all of that work will be done by and through the efforts of various individuals. As so much of the change then involves people, it is vital any integration process addresses the need to shift cultures and the behaviors of people.

Kotter and Cohen developed a seven-step process for implementing change that incorporates the need to address and capture the 'heart' as well as the minds of those being asked to change (see Figure 1 – page 157). Briefly those seven steps are:

1. Create a clear, inspiring and achievable vision of the future that speaks to the hearts as much, if not more so, than to the minds.
2. Leaders must build a sense of urgency for the change.
3. Identify, organize and mobilize a cadre of 'change leaders'
4. 'Change leaders' must be equipped and able to deliver concise and heartfelt messages to create trust in and among the wider 'audience'
5. Start 'busting barriers' whether those barriers are the current organizational structure, current systems or programs or actual day-to-day work practices and behaviors.
6. Set and achieve 'short-term wins' and widely advertise the achievement of such wins.
7. 'Victory' can only be claimed after the new behaviors, processes and practices are wholly embedded into the fabric of the day-to-day operations.

The steps, then involved in a successful integration process can be summarized as:

Step 1:

- Identify change champions among key individuals that can impact process safety,
- Engage these champions in developing stories and communication material that engages emotions as well as logic that will present a clear vision of the 'future',

- Then - Communicate, Communicate, Communicate and then – communicate some more!

<u>Step 2a – If using a top-down approach:</u>

- Appoint and charter integration teams of the following nature:
    - An 'Intolerable Risks' team, to identify short-term corrective actions addressing situations or risks the company deem are at an intolerable level
    - A Guiding team, with various responsibilities including having the authority to 'bust barriers' that become an impediment to the integration process moving forward therefore at least one executive or senior manager, should be a member of this team,
    - An Organizational team, with the objective of identifying synergies in and among the current process safety related structures as well as assuring any future organizational structure delivers an effective and efficient process safety program,
    - A Physical Equipment and Operational Process team, to identify gaps in the state of the current equipment and/or process and develop cost effective solutions to such variances.
    - A Process and Systems team, to identify and implement the critical process and systems, including IT systems, to support and help drive the desired approach to process safety,
    - A People team, to assure the necessary skills, competencies and culture exists to sustain an effective process safety program

<u>Step 2b – If using a bottom-up approach:</u>

- Appoint a team to coordinate, oversee and address issues beyond the scope of the various focus teams' authority,
- Prioritize the various activities and tasks to be addressed,
- Formally charter each focus group, and assure the charter contains a deliverable as well as a target deliverable date.

- Keep tasks 'short' ideally no longer than six months from start to completion and break large changes into shorter tasks where necessary,
- Widely communicate the success(es) of the focus teams to build and sustain momentum for the changes.

Step 3:

- Assure the vision of the desired state is clear and key individuals begin the process of modeling the new desired behaviors,
- As the process safety integration lead, be prepared to play a variety of roles but balance these against assuring the vision and strategy of the future continues to be communicated in clear simple motivational messages by key individuals,
- Assure sufficient time and resources are allocated to transitioning the changes to operations, do not allow the effort to achieve transitioning to be eroded,
- Make provision for audits of the planned changes to be undertaken after an agreed period of time such as a year or at most two years

Two alternatives for step 2 (mobilization and tasks of the actual integration teams) are discussed. This is in recognition of the fact some companies may prefer to take a 'top down' approach to the integration process while others may prefer more of a 'bottom-up' approach. Studies of the effectiveness of these two approaches have found a bottom-up approach has greater potential for success where there is a high level of agreement on both the general direction of the change as well as how to get there. On the other hand where there is a low level of agreement on these points the research suggests a top-down approach would be more successful. Which approach you wish to take should be actively discussed with the senior management integration team. It may well be they have decided on a general approach to integrating other programs, systems, etc. that could put your approach at tension with the one they have adopted.

When working through the integration process, a point to remember is that other parts of the organization, the systems, the human resources, the culture, etc. are going through an integration process as well. Changes made in those areas such as the implementation of a new computerized maintenance management system, changes to the incident and accident reporting systems and even what might seem to be changes in totally remote systems such as the time and expense reporting systems will impact the changes that will be made to the process safety programs. Understanding the scope of what other systems, programs or changes to the organizational structure are likely to occur is important and must be taken into consideration as part of your integration process.

Finally to confirm that the new changes have been effectively implanted into the newly integrated operations, the group or team responsible for operationally auditing organizational activities should be provided with material they can incorporate into their audit manuals and protocols of the changes. Discussions should be held with this group to determine whether it is advisable to arrange for a special audit of all the process related safety changes after a period of a year or at most possibly two years or to roll that into the normal set of scheduled operational audits.

CHAPTER 7 – M&A IN THE FUTURE

At the time this Guideline was being developed (i.e. late 2008 - early 2009) the world was going through a period of economic turmoil many were calling unprecedented while others were describing as falling just short of the 1929 Great Depression. The ability to obtain credit to finance a possible acquisition or merger was to say the least difficult. Yet even during that period, certain companies were going forward with M&A's. Dow Chemical closed on a deal to acquire Rohm and Haas in April 2009. A deal estimated at US$16.3 Billion.

However, of interest at a macro-economic level was that general reserves or supplies of monies were shifting from such areas as Europe and North America to new regions. For example, the Abu Dhabi Investment Trust was estimated to have assets valued at over US$800 Billion. As a result, even within this economic turmoil and period of tight credit potential funding for a merger or acquisition could be found. However it is likely the location(s) from where the actual financing is obtained may see a major shift in focus. That said, while all this is occurring at a 'macro' scale other than the volumes of such deals, the belief is PS issues currently and will continue to fall under the radar of such matters.

Changes in technology, demographics, socio-political factors and perhaps most important of all, shifts in norms, mores and cultures at the regional, country and local levels, etc. are more likely to impact process safety related matter in the future acquisition of companies.

Do we see process safety or major hazard safety management continuing to evolve and improve? Most definitely, yes. It may well be over the next few years depending on how the current 2008-09 economic turmoil plays-out we will see a plateauing in interest with respect to new regulations or enforcement actions. However, once we move through this period of economic turmoil, a period which is likely to coincide with some major demographic shifts as well, it is likely the professionals in the process safety field will have to undertake some fundamental rethinks of what is currently accepted practice. If we are correct in such projections and should the M&A activity 'heat-up' at the same time these demographic changes kick-in, the PS profession as a whole and especially those directly involved in merging and integrating what could be diverse programs together are likely to find their lives exciting.

THE APPENDICES

Three spreadsheets or tools were developed to assist with the tasks that lie in front of you.

The first is a checklist of potential process safety issues. This was developed to assist with building your own unique checklist of issues that should be examined in a potential M&A. As part of the due diligence you will need to examine these issues as input to your company's Vendor Due Diligence report if a Seller, or as part of the bid you will submit if a Buyer. After the deal closes, you will next need to establish a baseline of the current state of process safety efforts in the newly acquired assets vs. the desired state senior management wish achieved. Again we want to stress this checklist is there to assist you in developing your own checklist to address the unique set of circumstances associated with any one particular acquisition or merger. While it covers approximately thirty-eight pages if printed out in landscape format, even here such a generic checklist cannot address or cater for all the factors or issues that will be associated with your company and the assets being considered for divestment or acquisition.

The second and third spreadsheets were developed to help you build an integration plan and estimate the potential costs associated with that plan. The exemplar budget caters for the mobilization of multiple teams to address the various activities as laid out in the exemplar plan.

These two spreadsheets are adaptations from an actual integration process that entailed two very large organizations. When the two organizations' assets were combined, their total market capitalization was estimated at approx US$38 Billion (in 2000-01 dollars). Further, in doing the 'gap analysis' it was found the two organizations had taken fundamentally different approaches to the manner by which it approached HSE in general. This approach cascaded through their process safety programs. There were other significant differences in the manner by which they approached process safety related matters, as well. Finally, the new 'parent' took the decision that the new assets would be fully integrated together using the best of both. Hence the reason the spreadsheets are as comprehensive as they are. They should, however, provide you with a good starting point more to delete issues or specific activities than necessarily have to identify and create such points.

1

AN OVERVIEW OF PROCESS SAFETY

"To know is to survive and to ignore fundamentals is to court disaster."
H.H. Fawcett[1]

1.0 COURTNEY'S STORY – CONTINUED

If you have not read the preface, you will not have been introduced to Courtney. If you have a few minutes take the time to go back and become familiar with Courtney. Her story will continue at the beginning of each chapter.

Carla the VP of Compliance and Regulatory Affairs, and Courtney's boss continued to explain the business aspects of the proposed acquisition of White Hot Chemicals. However Courtney's mind continued to focus on what Carla had just said, that as part of Courtney's responsibilities they wanted her to assess the state of White Hot's process safety programs as well as all other HSE related matters. In their US operations Bland's process safety matters were handled by the Asset and Operations Integrity (A&OI) group, not HSE. When Courtney raised this point, her boss explained because of the turn-around at Bland's Pasadena, TX site and the expansion at Port Arthur, TX the A&OI group didn't have anyone they could assign to the due diligence team. In fact, one of the reasons Courtney was assigned to the due diligence team was Courtney oversaw the major hazards programs of Bland's Fawley, UK operations and was the HSE director there for three years. That plus Carla noted that White Hot's Ternuezen facility was one of the potential real 'gems' in the acquisition and they felt Courtney's previous experience with major hazard legislation in Europe would be a real plus. That is why they had every confidence Courtney could handle the process safety as well as all the other HSE related issues. Courtney thought to herself – "I wish I had that same level of confidence!"

1.1 WHY THIS GUIDELINE?

You have probably heard the adage that only 10% of an iceberg is visible above the waterline, the other 90% being submerged or hidden from view. The development and publication of this Guideline was based on similar findings of Center for Chemical Process Safety (CCPS) member companies – namely many process safety issues were found only after they had purchased a set of assets, facilities or operations. In other words, prior to the purchase and despite doing a due diligence review of the site or business, these process safety issues were 'below the waterline' only to be discovered later after the deal closed.

Figure 1. The Iceberg principle

This situation is not new to the health, safety and environmental community. Often in the past environmental legacy issues were not investigated prior to purchasing a site or business. Afterwards and with changing legislation many companies found they had 'inherited' significant environmental liabilities with the purchase. Sometimes these liabilities required investments of tens of millions of dollars in order to remediate or correct conditions left by the previous owners. Today, investigating environmental matters forms an integral part of the due diligence process when purchasing a facility that stores, processes, manufactures or handles hazardous materials or chemicals. The same holds for certain health issues such as those associated with asbestos. However, this practice has yet to become standard or routine when it comes to identifying and evaluating process safety or major accident hazard issues as an integral part of undertaking a due diligence investigation of a potential acquisition.

While there are a number of explanations for this, perhaps the primary reason is management not recognizing the potential costs that can accompany correcting or rectifying gaps in an operations approach to process safety or major hazard risks. Against this however, a number of the member companies of the Center for Chemical Process Safety investigated the costs they were incurring to rectify process safety issues found after they had closed on an acquisition. Arising out of their investigation they found it was common that another **ten to thirty percent** of the initial purchase price was being expended rectifying such issues. And these were not rare or extreme cases, rather they represented the average costs they were experiencing. Why so much? Examples included:

- Costs of between $30-40 Million just to build and equip a new control room, when studies found the current control room too near to major inventories of flammable materials exposing occupants to an unacceptable level of risk,
- A variety of issues with plant, equipment, control systems, etc. where on studying the issues further, the company found

the costs of 'retro-engineering' corrections amounted to almost 100 times the costs of engineering a new facility.

- Following the Texas City refinery explosion and fires, new requirements for relief and blow-down systems and the siting of occupied buildings or structures are likely to require considerable rectification and upgrading of these systems in light of changes to standards, codes and recommended practices.

In summary, one CCPS member noted:

> *"Rectifying occupational health and safety issues generally runs us hundreds of thousands of dollars. Rectifying process safety issues has been costing us in the tens of millions of dollars. It was not until we were able to prove and present these hard facts to management that process safety became an integral part of our due diligence process."*

The above is 'lesson one' of this Guideline – i.e. the need to get process safety issues onto the agenda of your company's due diligence investigations or studies.

Lesson two of this Guideline is also born out of hard experience. That integrating or merging the process safety approach or programs of two companies can be as equally challenging. This has been found even where a company acquiring a set of assets has itself, operated similar facilities or processes for a period of time. Case studies and discussions of these issues can be found later in sections 2.0 and 2.5 of Chapter 2.

As a result then, this Guideline was developed to:

- Pass some of these 'learnings' on to others to help you make a strong case to your management that process safety issues need to identified and investigated thoroughly in the M&A due diligence process,
- To provide assistance on the various types of process safety or major hazard issues that should be investigated, and

- Provide assistance with planning the process for integrating together the process programs of the two companies after the deal has closed.

1.2 UNDERSTANDING THE BASICS

This first chapter introduces basic concepts related to chemical process safety including an understanding of risk versus hazards. In addition, it is important for every organization to understand the key differences between occupational safety and chemical process safety.

When one hears the word "safety", the general reaction is to think of personal injuries and/or minor accidents, such as cuts, bruises, falls, motor vehicle accidents, or muscle sprains and strains. All of these safety issues can be very serious. However this type of safety incident is typically isolated to a single individual and is generally called occupational safety. When a hazardous materials related incident occurs it often impacts more than one person in addition to property damage and interrupting business flow(s) and possibly off site impacts as well. Process safety hazards can give rise to major accidents involving the release of potentially dangerous materials, the release of energy (such as fires and explosions), or both. Process safety incidents can result in multiple injuries and fatalities, as well as substantial economic, property, environmental damage, and community impact.

The current working definition for Process Safety as used by the Center for Chemical Process Safety is:

> *Process Safety – is a disciplined framework for managing the integrity of operating systems and processes handling hazardous substances by applying good design principles, engineering and operating practices. It deals with the prevention and control of incidents that have the potential to*

release hazardous materials or energy. Such incidents can cause toxic effects, fire or explosion and could ultimately result in serious injuries, property damage, lost production and environmental impact.

> *1. A discipline that focuses on the prevention and mitigation of fires, explosions, and accidental chemical releases at process facilities. Excludes classic worker health and safety issues involving working surfaces, ladders, protective equipment, etc.*
>
> *2. A discipline that focuses on the prevention of fires, explosions, and accidental chemical releases at chemical process facilities*

1.3 HAZARD VERSUS RISK – IS THERE A DIFFERENCE?

Two other terms the reader should be familiar with, as they have very distinctly different meanings, are "risk" and "hazard". Both terms are defined and discussed in numerous CCPS guidelines and materials. Below are two definitions extracted from such other material:

> *Hazard – A chemical or physical condition that has the potential for causing damage to people, property or the environment. A hazard is intrinsic to the material or to its conditions of storage or use. With respect to chemicals, "hazard" may include toxicity (acute or chronic), flammability, corrosivity or reactivity.*
>
> *Risk – A measure of potential loss (for example, human injury, environmental insult, economic penalty) in terms of the magnitude of the loss and the likelihood that the loss will occur.*

To provide an everyday example that might help explain these terms, the risk of injury in a motor vehicle accident is much higher than the risk of being fatally injured in an airplane accident. According to statistics published by the U.S. National Transportation Safety Board, the risk of being fatally injured in an aircraft accident is 0.0002 fatal injuries per one million miles flown as compared to the risk of being fatally injured in a traffic accident, which is 1.4 fatal injuries per one million miles driven or traveled. In other words within the United States, you are 7,000 times more likely to be fatally injured for every one million miles traveled in a car versus in an airplane. However, in an airplane crash more individuals are likely to be injured in a single accident. The likelihood of an event occurring causing injury to the passengers in a motor vehicle is extremely high relative to the likelihood of an airline crash causing injury to its passengers. This is an important concept when discussing chemical hazards and process safety. The public is willing to accept a large number of accidents that involve a few individuals as occur in motor vehicle accidents, but does not tolerate a single event that injures many individuals.

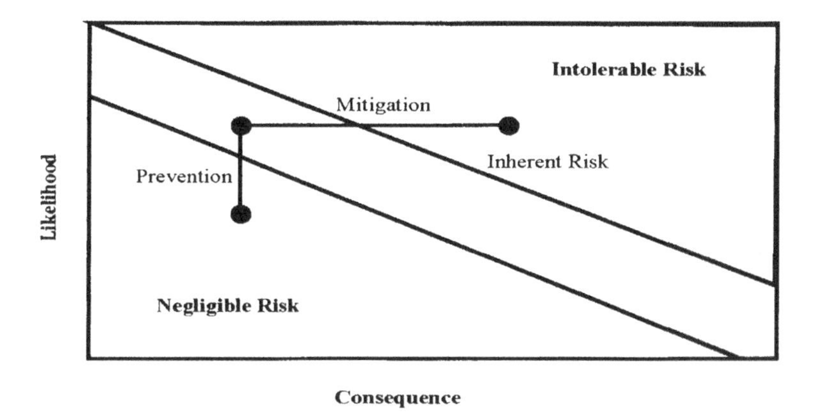

Figure 2. Risk Tolerability

Figure 2 presents the various elements that make-up risk, namely the likelihood a hazard or event might occur coupled with the potential consequences that could arise. The Figure also demonstrates that as the consequences become more severe (e.g. multiple injuries or fatalities, major environmental damage, business interruption) the acceptance of those consequences becomes decreases. The same is true for the likelihood of a particular class of accident occurring. Simply stated, when incidents or accidents increase in frequency or their consequences become more severe the acceptance or tolerability of those accidents decreases. The Figure also shows the impact various risk control measures might have on any one hazard or risk. Risk controls that help mitigate the consequences of an accident move the 'acceptability' of those consequences to the left in the chart. Preventive risk control measures on the other hand act to reduce the likelihood of a particular risk from occurring thus moving their acceptability down in this figure. Most risk control programs will work to identify the optimal combination of measures taken to reduce both the likelihood as well ways to reduce the consequences with the ideal of moving the risks into the lower left hand corner (i.e. to a tolerable if not negligible level).

Process safety accidents can impact numerous individuals both on and offsite. These low likelihood, but high consequence events cause great unease in and amongst the public. As a result, considerable resources are spent to prevent their occurrence. There are numerous techniques for assessing process hazards in new and existing facilities. A similar approach needs to be used when assessing process safety during the merger and acquisition process.

1.4 GOOD INJURY RATE DOES NOT EQUAL GOOD PROCESS SAFETY PERFORMANCE

When an acquisition or merger is being considered, a key performance metric typically requested is one that is calculated based on employee injuries and hours worked. Another common key performance metric is one associated with inspections or violations noted by governing agencies. Neither of these metrics though is appropriate for measuring a company or facility's performance as it relates to process safety.

An example of this is based on findings in the U.S. Chemical Hazards and Safety Investigation Board's (i.e. CSB) investigation report on the BP Texas City Refinery explosion. In the CSB report the investigation panel noted BP had emphasized personal safety, and had in fact been very successful in injury reduction over the last decade. However BP did not have process safety performance metrics in place. As process safety expert Trevor Kletz notes,

> *Personal safety metrics are important to track low-consequence, high-probability incidents, but are not a good indicator of process safety performance. The lost time rate is not a measure of process safety. An emphasis on personal safety statistics can lead companies to lose sight of deteriorating process safety performance. Process safety metrics provide important information on the effectiveness of safety systems, and an early warning of impending catastrophic failure.*

CCPS has recently issued a publication defining process safety leading and lagging metrics, Process Safety Leading and Lagging Indicators and it is recommended these be used rather than personal injury statistics.[2]

In summary, the injury performance at a facility, which handles highly hazardous compounds, may not be indicative of process safety performance. However, a consistently poor injury rate can be viewed as an indicator of the level of management commitment to operating a facility safely.

1.5 UNDERSTAND THE HAZARDS OF CHEMICALS HANDLED ON SITE

When considering a merger with or acquiring a company that handles hazardous chemicals, a key first step is to identify what chemicals are handled on site, the quantities stored and how they are processed. Once it is determined there is a process hazard at a particular facility, the merger and acquisition team must examine the hazard identification techniques employed in the past, but more importantly the preventative measures currently in place.

There is no prescriptive way to identify high hazard chemicals. Every chemical can potentially be hazardous. There have been numerous efforts to make lists of highly hazardous compounds. European countries, Asia, the United States as well as many other countries have developed high hazard chemical lists. The chemicals on those lists are selected for a variety of reasons. Factors considered in developing various high hazard lists include:

- Inherent toxicity, for example - chlorine and anhydrous ammonia;
- Flammability, for example - ethanol, methanol or butane;
- Reactivity, for example - peracetic acid;
- Environmental impact, for example - chloroform;

There are even lists of highly hazardous substances, based on the potential for a chemical or substance to be used as a weapon for terrorists or sabotage. Those lists may include highly potent active pharmaceutical ingredients or items that would be utilized to make weapons of mass destruction. There are many substances on these lists because they pose multiple hazards. Ethylene oxide is a prime example of a substance that is highly toxic, highly flammable, environmentally sensitive and reactive.

With such a wide variety of potential hazards, a starting point for hazard identification is needed. One option is to obtain a master list of all chemicals handled on site and compare the list to the various lists of hazardous chemicals developed as noted above. Keep in mind when reviewing the chemicals handled onsite that intermediates and byproducts need to be considered in addition to the raw materials and the final products. Each stage of the process must be considered to identify all highly hazardous substances handled in a process.

Each facility, which handles highly hazardous chemicals (HHCs) should have an understanding of which items are covered by process safety or major hazard accident regulations as well as any applicable industry codes and standards. Table 1 provides a summary of international process safety related standards, codes, and regulations. While extensive it is not an exhaustive or all-inclusive list of relevant codes, standards and practices.

Knowledge in this field of expertise is required to fully comprehend the requirements associated with a facility that has process safety issues. Keep in mind that these lists or regulatory requirements are only a starting point. If a facility does not handle one of the substances on one of the lists or does not exceed the threshold quantity listed for a particular chemical that in itself, does not mean a process safety hazard is not present. A prime example of this is represented by an incident that occurred in 2003 at Catalyst Systems Inc. This incident involved only 200 lbs. of benzoyl peroxide.

On January 2, 2003, a vacuum dryer holding nearly 200 pounds of benzoyl peroxide exploded at the Catalyst Systems Inc. production facility in Gnadenhutten, Ohio. Employees were in the process of drying granular benzoyl peroxide, which is unstable at high concentrations, when the explosion occurred.

Once there is an understanding of the chemicals handled at the facility, individuals with a process safety background should be added to the M&A assessment team.

Table 1. Examples and Sources of Process Safety Related Standards, Codes, Regulations and Laws [3]

Various sources of process safety related information
Voluntary Industry Standards
• American Petroleum Institute Recommended Practices[4] • American Chemistry Council Responsible Care® Management System and RC 14001[5] • ISO 14001 – Environmental Management System[6] • OHSAS 18001 – International Occupational Health and Safety Management System[7] • Organization for Economic Cooperation and Development – Guiding Principles on Chemical Accident Prevention, Preparedness, and Response, 2003[8]
Consensus Codes
• American National Standards Institute[9] • American Petroleum Institute[4] • American Society of Mechanical Engineers[10] • The Chlorine Institute[11] • The Instrumentation, Systems, and Automation Society / International Electrotechnical Commission[12] • National Fire Protection Association[13]
U.S. Federal, State, and Local Laws and Regulations
• U.S. OSHA – Process Safety Management Standard(29 CFR 1910.119)[14] • U.S. Occupational Safety and Health Act – General Duty Clause, Section 5(a)(1)[15] • U.S. EPA – Risk Management Program Regulation (40 CFR 68)[16] • Clean Air Act – General Duty Requirements, Section 112(r)(1)[17] • California Risk Management and Prevention Program[18] • New Jersey Toxic Catastrophe Prevention Act[19] • Contra Costa County Industrial Safety Ordinance[20] • Delaware Extremely Hazardous Substances Risk Management Act[21] • Nevada Chemical Accident Prevention Program[22]

Various sources of process safety related information
International Laws and Regulations
• Australian National Standard for the Control of Major Hazard Facilities[23] • Canadian Environmental Protection Agency – Environmental Emergency Planning, CEPA, 1999 (section 200)[24] • European Commission Seveso II Directive[25] • Korean OSHA PSM Standard[26] • Malaysia – Department of Occupational Safety and Health (DOSH) Ministry of Human Resources Malaysia, Section 16 of Act 514[27] • Mexican Integral Security and Environmental Management System (SIASPA)[28] • United Kingdom, Health and Safety Executive COMAH Regulations[29]

Highly hazardous chemicals need to be properly controlled and contained at all times. The basic design of the equipment is critical to ensure proper containment is achieved. If the equipment is not of a proper design or is not "fit-for-purpose" the potential for a chemical release is inevitable. If the equipment is not well maintained, the results can be equally catastrophic.

There are numerous industry groups that provide guidance on process safety for specific chemicals. For example, chlorine is a common chemical that can be found in multiple locations across the U.S. Chlorine suppliers and their clients also have a history of incidents and continue to learn and establish new codes and standards that are designed to prevent a release. The Chlorine Institute is a key resource for these efforts. Due diligence teams would need to understand if prospective facilities that handle chlorine have followed the design and engineering codes established by the Chlorine Institute. Other consensus codes that address process safety are listed in the previously noted Table 1.

1.6 DON'T FORGET ABOUT THE DUST EXPLOSION HAZARD

A dust explosion hazard is present with many common solids. Quite often these solids may not historically have been considered a high hazard. Dust explosions have occurred for years, particularly in the grain handling business sectors. A recent incident should be a key reminder of the potential catastrophic hazard associated with a dust explosion. An explosion that occurred in February of 2008 was a clear indication of the power of any combustible solid material when suspended in air. On February 7, 2008 an incident at an Imperial Sugar refinery resulted in 13 fatalities (on August 22, 2008 a 14th fatality occurred after months in critical condition) and a total of 40 individuals injured.[30] The explosion was believed to have been the result of an initial 'minor' explosion followed by a larger explosion. The first minor explosion resulted in dust dislodging from various surfaces. This dust also became suspended in air resulting in a second catastrophic explosion. This scenario of a small primary explosion followed by a catastrophic secondary explosion is stereotypical of a dust hazard. The Chemical Safety Board completed a comprehensive investigation on the history of dust explosions. Their investigation was initiated after three major explosions in 2003 resulted in 14 deaths. As part of their investigation the CSB investigators researched the history of dust fires and explosions from 1980 to 2005, and identified 281 major combustible dust incidents that killed 119 workers, injured 718 others, and destroyed many of the industrial facilities.[31] The potential catastrophic result of a dust explosion necessitates the completion of a thorough assessment of this particular process safety hazard as part of the M&A process.

Many materials when suspended as a dust in air can lead to explosions. As described above sugar is one. Grain dusts have led to a number of explosions in grain handling silos. Coal dusts, of course, as well as wood dust(s) and many different metal dusts such as aluminum have also been the source of airborne-suspended dust explosions. Sites, facilities or processes with the potential to generate such dust hazards, must be evaluated to assure controls are in place that manage the risks of an accident with potentially catastrophic consequences.

1.7 UNIQUE CONSIDERATIONS AT FACILITIES THAT HANDLE HHCS

Chapter 4 provides further detail on the issues and activities that should be carried out to assess the state of the current process safety programs as part of a due diligence investigation. However, to assist with establishing a fundamental knowledge or understanding of the management of process safety issues, a few areas aspects of an effective approach to PS are introduced here.

Process safety is an extremely broad and technical topic in itself. Just developing a basic knowledge and understanding of which codes and practices are applied to various process safety hazards is a monumental task. The codes are ever changing as the industry experiences incidents and review panels modify standards based on new knowledge. New publications are issued frequently. For example, one design standard mostly unique to the chemical, refinery and related industries is facility siting. Facility siting involves proper spacing and organization of a chemical processing complex between on site process units themselves as well as relative to its neighbor's. Industry standards have been changed recently with new requirements for locating portable buildings (API 753 – "Management of Hazards Associated with Location of Process Plant Portable Buildings" issued in June 2007[32]). These changes were made to ensure such facilities and their occupants are protected against toxic releases, fires and explosions. Facility siting assessments not only look at spacing, but also assess occupancy load and the construction of the building in question.

Facility siting involves examining potential offsite impacts as well. The scenario examined here is the proximity of buildings in the surrounding neighborhood or near to the fence line of an industrial facility that handles HHCs. This scenario is a very difficult risk to mitigate after the fact - i.e. after construction. Noteworthy major accidents or catastrophes where neighboring residential buildings were severely damaged or members of the general public were seriously or fatally injured include:

- 1974 – The explosion at the Nypro chemical works at Flixborough, UK
- 1976 – The release of TCDD (dioxin) from the ICESMA chemical works near Seveso, Italy
- 1984 – The release of methyl isocyanate at the Union Carbide Bhopal, India chemical works,
- 2005 – The explosion and fires at the Buncefield, UK fuel storage and transfer depot.

More recently the potential magnitude of offsite impacts can be found by viewing a video available on the United State's - Chemical Safety and Hazard Investigation Board's (CSB) website entitled "A Blast Wave in Danvers"[33] and the associated incident investigation report CAI/Arnel Chemical Plant Explosion Danvers MA. This video demonstrates what results when a process safety incident occurs at a facility in close proximity to a residential area. These issues can easily increase the risk of a potential buyer when seeking to acquire a new facility or merge with a company that has multiple sites some of which are near to residential areas.

The above incidents also raise the need to investigate and evaluate facilities adjacent to a potential acquisition site or operation. This is especially true where an asset is located in the middle of an industrial complex that comprises multiple facilities storing, processing or handling highly hazardous chemicals. You may well be faced with the situation where the only thing separating the site or assets being considered from an adjacent facility are a few 'tens of feet' or meters and a chain link fence. Further, companies are acquiring specialist facilities or operations that exist in the heart of a major chemical or petrochemical complex where the whole function of that facility is providing or feeding the units around it with power, steam or possibly a unique process stream. Such acquisitions are becoming more commonplace. These purchases involve unique issues that will need to be investigated and evaluated from a process safety standpoint as part of the due diligence process. The potential liabilities with such purchases must be well understood to ensure proper controls are in place and whether following the purchase one or multiple parties will assume the risks and associated liabilities.

While the physical features of a prospective facility are critical aspects that influence the process safety risks, the management systems that are in place play a dominant role in determining how well those risks are controlled. The chemical and petroleum industry is one where good management systems are essential to the safe operation of the various processing units and operating facilities. Examples of management system requirements include management of change, incident investigations, employee participation, control of hot work and so on. Each of these systems is a key aspect of the risk controls that will need to be examined in the due diligence portion of the merger or acquisition. Further, assuring these systems remain robust will be essential when going through the changes of a corporate re-structuring once the merger or acquisition occurs. In his book *"Still Going Wrong"* published in 2003 Professor Kletz noted that companies at that time may not be evaluating the impact of the full range of organizational changes on process safety or major hazard management programs. He went on to recommend the impact from changes such as outsourcing, major re-organizations, mergers and 'downsizing' should all be assessed by Management of Change programs.[34]

1.8 RESOURCES FOR PROCESS SAFETY

A key resource in which to find further information on process safety issues is the library of books published over the last several years by the Center for Chemical Process Safety (CCPS). The Center identifies and addresses process safety needs within the chemical, pharmaceutical, and petroleum industries. CCPS offers over ninety titles providing current guidance for all those that produce, store, and handle flammable, explosive and reactive materials. The CCPS library of guidelines, addresses the full range of process safety challenges, from inherently safer process design, to hazard evaluation and safe design, to advanced process safety management practices, to incident investigation.

There are multiple other resources available to develop a basic knowledge of process safety (see Table 1). The CSB is an independent federal agency charged with investigating industrial chemical accidents. The "Blast Wave in Danvers" video, mentioned earlier in this Chapter is just one of various videos and associated accident reports available on CSB's website. This website is a great resource for any individual who wants to understand the risks associated with process safety incidents. The videos on this website are of high quality. They would serve as a great educational tool for a due diligence team associated with an acquisition or merger in the chemical industry.

A basic textbook on process safety is entitled - "Chemical Process Safety: Fundamentals with Applications".[35] It provides a comprehensive introduction to the essential technical fundamentals of chemical process safety. Its emphasis on fundamentals is intended to help both experts in the field of process safety as well as those desiring to obtain a basic understanding of process safety.

In addition to the CCPS the American Petroleum Institute (API) is a valuable resource for both the petroleum and chemical industries. As with the CCPS, API has developed a wide range of industry standards, recommended practices and guidelines as well as many other resources directed toward the safe production, refining and transport of petroleum based products.

Another key organization is the National Fire Protection Association (NFPA). Many NFPA codes have been adopted as regulations at the state and national level. Other NFPA codes are recognized as establishing good industry practice or Recognized and Generally Accepted Good Engineering Practice (RAGAGEP) for protection against various hazards.

These are only a few of the many resources on process safety available. However, by far the most important resource in a merger or acquisition is the technical experience and expertise of staff or consultants familiar with the chemicals being handled, and the proper safeguards and designs that are appropriate for the hazards present. Even that expertise though may be facing their first acquitision of a new company and the ensuing process for merging two possibly disparate approaches to process safety or major accident hazard prevention together. The following chapters seek then not only to assist an individual new to process safety as well as the individual who is well versed in process safety but new to an acquisiton or merger. The objective of each chapter is to help such individuals and their organizations factor process safety decisions into the various stages of an M&A in a timely and effective manner.

2

The Merger and Acquisition Process

Of the 55 refineries closed in America in the last 10 years, they were all closed for economic reasons, mostly oil company mergers. Not a single one was closed for environmental purposes or objections.
Peter DeFazio

2.0 COURTNEY'S STORY - CONTINUED

Courtney's PDA buzzed as she was walking back to her office after a meeting with her boss, Carla. When she opened the message it noted the kick-off meeting for the due diligence team had been rescheduled to start at 2.00PM. The flights of a couple of individuals who were to attend were delayed.

When Courtney did get back to her office she had a chance to sit back in her chair. Her mind again turned to the fact that she had never been directly involved in an M&A. Her first thoughts were – "OK, everyone in our industry has heard the terms mergers and acquisition or M&A, due diligence, integration, but what do they really mean and what do they actually involve? Is a merger the same as an acquisition? What are the responsibilities of a due diligence team and what does a 'due diligence' team really do?"

She started to search for these terms on the Internet and was still in the midst of her searches when the meeting reminder popped-up on Courtney's computer screen telling her she only had an hour to drive cross-town to catch the 2.00PM start.

2.1 CHANGING WORLD OF CORPORATE PROFILES

The world of mergers and acquisitions has become standard headline news in the financial markets. As a result, the chemical and petroleum industry's business portfolios are ever evolving. Organizations are now more than ever merging, acquiring, divesting, splitting, and even spinning off new companies. Many employees have worked for an employer that has been bought out, merged or spun off under a new name possibly several times. Quite often this happens in just a few short years. The profile of corporations and the core businesses within the corporations is an extremely dynamic process in this current world market. This guideline is targeted at those companies that are acquiring or divesting chemical, petrochemical and petroleum process facilities through mergers acquisitions or disposals. According to Chemical Weekly over half of all mergers and acquisitions in the chemical industry occur in the U.S.[36] The environmental, safety and risk considerations related to companies or facilities that handle highly hazardous chemicals add a layer of complexity to merger and acquisition considerations that are not typically a concern in the nonchemical industry business sectors. A key element that needs to be managed through every one of these mergers, splits, acquisitions and spin offs is process safety.

At the time of the writing of this Guideline, the European Union remains the largest producer of chemical products followed by the Asia-Pacific region and then the USA. The traditional dominance of chemical production by these countries is being challenged by changes in feedstock availability and price, labor cost, energy cost, differential rates of economic growth and environmental pressures. Instrumental in the changing structure of the global chemical industry has been the growth in countries such as Saudi-Arabia, Kuwait, China, India, and Korea among many others. Not only are stereotypical chemical and refinery industries acquiring and merging in the world market to make new organizations, but nonchemical manufacturers are also venturing into the chemical industry as it has recently proven to be a profitable investment. This new world market and diversity within corporate profiles includes multiple business strategies. These diversified strategies have led to mergers and acquisitions at an ever-increasing rate.

> *Early 2003 was the turning point. Chemical companies with an improved balance sheet became more active in M&A deals. The chemical industry's interests in M&A are rising because they need a way to increase scale. After years of holding off and being internally focused chemical companies have started to look externally to find ways of improving their balance sheets and creating value*[37].

In addition, the investment community and many corporations now make it their business to drive mergers and acquisitions. The chemical industry attracts private equity firms.

> *The traditional model is for private equity firms to either strip costs out of their investments or bundle them with similar or complementary businesses before selling them. The average life of an investment up to the point of sale is three and a half to five years.*[1]

2.2 OVERVIEW OF THE M&A PROCESS

The Merger and Acquisition (M&A) process as outlined in Figure 1 is fairly simple, but the details can have a major impact on the outcome of the process. Surprises late in the negotiations can derail them altogether or at the very least lengthen the sales process.

Figure 1. Overview of the M&A Process

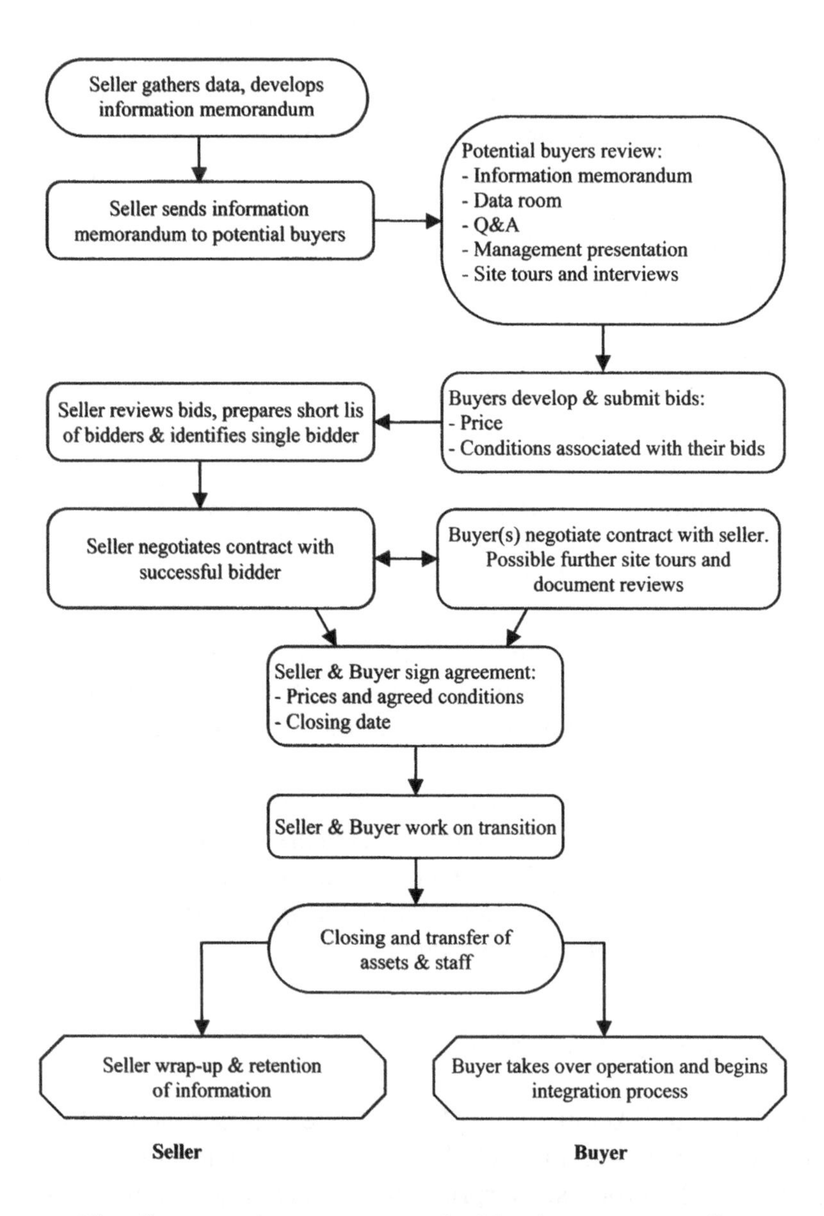

The first step is a company decides it wants to sell an asset, which is usually an internal decision by the management team. Once the decision is made, the seller then gathers information relative to the asset so it can be marketed. The data and documents are then put into a data room and used to develop a document (often called an Information Memorandum) distributed to companies that have shown interest in buying the asset. Following the signing of a confidentiality agreement prospective buyers or bidders are then given access to a data room and the opportunity to participate in a data room Question and Answer (Q&A) process. Bidders are usually offered management presentation meetings with the seller's management. These meetings may or may not include a brief tour of the assets.

After the bidders have had an opportunity to review the information, they will conduct a business valuation of the assets then make a bid. This bid usually includes a price and conditions on which they would acquire the asset. A bid date is usually set by the seller, at which time all the bids must be formally submitted. After the bids are received, the seller reviews them, compares the purchase prices and the various conditions that the bidders have specified for the purchase

The seller conducts a business valuation of the bids, and selects one (or more) potential buyer(s). If none of the bids are acceptable, the seller can negotiate with certain of the bidders for revised bids, or the seller can simply cancel the sales process.

Once the buyer is selected the seller and buyer enter into detailed contract negotiations on the final conditions of the sale and contract language. Also during this time the buyer can conduct an on-site due diligence for further discovery. If during the site due diligence, a buyer discovers a major issue not disclosed in the data room prior to the bid, the buyer may want to change the bid price and/or the conditions of sale. For this reason, all major issues should be disclosed in the data room, and from the HSE/PS point of view, the Vendor Due Diligence (VDD) report. At the end of these negotiations a formal agreement is signed, which includes the price, all the conditions of sale and a closing date.

The closing date is usually a month or more from the date of signing to allow for any regulatory approvals for the sale to be processed and approved. During this time, the buyer and seller management teams work on the transition items that must be accomplished prior to or on the closing date. On the closing date, ownership officially transfers from one company to another, so the personnel, permits to operate, support systems, management systems, etc. need to transfer or change to the new owner at that time or shortly thereafter.

2.3 SCALABILITY (BIG/SMALL; SINGLE SITE VERSE MULTIPLE SITE DEALS)

M&A deals can range from very large mega-mergers of two multinational companies to the small, such as a sale of one service station. No matter what the size of the deal, the process is similar but the complexity and the size of the due diligence process and required team will be different. The use of a checklist such as the one included in the Appendix allows for multiple teams to work concurrently in different locations and deliver their findings in a consistent manner.

For large single site M&A projects there will likely be one person on the due diligence team with responsibilities for process safety, unless it is a high risk business with multiple non-standard high risk operating units. Though it may appear daunting at first, one person can do a very thorough due diligence of a facility by attending the data room, then spending approximately three to five days in a facility interviewing key personnel, conducting site tours and reviewing additional information while engaging in questions and answers with facility personnel.

If the project involves multiple large sites, there will likely be multiple teams with one process safety person on each team. One person can participate on more than one team if the site visits are not concurrent. Where the project is a divestment and the teams are preparing the Vendor Due Diligence reports, ample time must be scheduled to write up the findings.

For smaller sites, the process is the same for a large site, except there will be fewer key staff to interview, potentially fewer documents to read, a smaller site to tour and probably a shorter report. Also a facility visit will likely be from a few hours to one or two days. If the project involves multiple small sites, multiple teams using the same checklist can be used to expedite the process.

An acquisition may involve a large number of similar single sites such as the sale of a gasoline retail market of 100 service stations, the sale of an oil field with 150 wells, the sale of a distribution system with twenty-five warehouses, etc. In cases like these, managing the data can be one of the major issues relating to the due diligence activities. One way to address this issue is to develop a standard data table report that can be filled in from the records review and site visits. It may also be sufficient to only review a cross sample of similar facilities when or where they are considered low risk operations or in low risk areas.

2.4 KEY TERMS AND CONCEPTS

This Guideline uses the terms "merger and acquisition" and "M&A" process as a shorthand reference to a plethora of transactional structures only lawyers can conceive. But in the end, the net result is that processing plants, individual process units, manufacturing plants, transportation facilities, and other complex facilities change hands to a new owner and/or operator. These include: mergers, acquisitions, joint venture formations, divestments, decapitalizations, Initial Public Offerings (IPOs), asset swaps, etc. and contractual agreements including the transfer of contracts to supply goods and materials and toll processing agreements.

Acquisition – is the purchase of assets or equity interest in a company and may in either case include the transfer of the operating staff from a seller to a buyer.

Decapitalization – is the sale of the assets of a business, where the brand or operation of the facility remains with the seller. This is often the case where the selling company wishes to reduce its capital base whilst maintaining its operations and product base. Decapitalizations are common in retail gasoline markets where a branded major sells its stations to wholesalers or dealers while keeping the major's brand on the station. This is similar to a divestment.

Divestment – is the sale of assets or equity interest in a business entity and may include the transfer of the operating staff from the seller to a buyer

Due Diligence – is a term used for a number of concepts describing the performance of an investigation or assessment of a business or of certain assets and may address one or all of accounting, legal, engineering, environmental, and other analyses to evaluate the assets, liabilities, and potential liabilities of a business. The term commonly applies to *voluntary investigations*. A common example of due diligence in various industries is the process through which a potential acquirer evaluates a target company or its assets for acquisition. It is also prudent for the seller to conduct a due diligence investigation to identify material issues or risks associated with the assets that will be sold prior to offering them for sale - this is often referred to as *Vendor Due Diligence.*

Initial Public Offering (IPO) – An initial public offering is when a parent company packages one or multiple assets into a new company whose stock will be offered for sale to the public. This is similar to the divestment process in that a divestment type due diligence must be done on the assets and staff. The business team will use this information to assess the value of the assets and prepare a summary document for potential significant buyers of the stock.

Integration – For the purpose of this Guideline, integration is the process of planning and implementing various activities to merge together the operations, systems and staff of the newly acquired assets in accordance with the desires and expectations as established by executive management.

Joint Venture – A joint venture (JV) is the combining of assets and/or cash into a new company that is an independent legal entity from the companies forming the JV. JVs are usually managed by a board of directors comprised of individuals from the management ranks of each of the partners, with the representation of each company based on the asset and monetary value that each company contributes to the JV.

JV participation by the partners includes those with a majority partner (greater than 50% ownership or the largest ownership share of multiple owners) that would likely have significant control over the JV operation. Also those with equal partners (50/50 ownership) where each partner has equal control/influence but neither partner can mandate a course of action. A third form where a minority partner (less than 50% ownership) would likely have influence over the operation unless control is established in the JV formation contracts on through to a minor partner with less than about 10% interest. In these cases the minority partner(s) would act more like an investor(s).

Depending on the nature of the JV, the facilities can be operated by one of the partners under an operating agreement. In these cases the facility personnel usually remain employees of the operating company. Separate to this is the case where the plant or facilities are to be operated by the JV. Here the existing staff is transferred to the new company or various combinations of operating agreements are developed and entered into by the JV partners. How the JV is to be operated should be established early in the negotiation process, as this will have a major impact on due diligence activities and integration plans.

A JV can be thought of as a divestment of one partner's assets and an acquisition of the other partner's assets. Which assets one views as being divested or acquired is dependent on which partner you represent. Based on the value of the JV and the percentage contribution level of a company, levels of significance need to be established early in the process to prevent due diligence teams from concentrating on insignificant issues.

Materiality & Material – in the context of an M&A is the measure of the significance or effect that the presence or absence of an item or issue may have on the transaction. Materiality can be judged in terms of its impact, value, or risk with regards to compliance, liabilities or associated costs. It is not uncommon to assign a "materiality threshold" in dollar terms to set a threshold of individual or aggregate items to consider during due diligence. A rule of thumb is to use around 10% of the deal value as the materiality threshold. It is a good idea to consult the business or legal team to help identify an appropriate materiality threshold to use.

Merger – is technically the combination of two business entities where typically one entity survives and the other comes to an end, but with the practical result that the assets, staffs and managements (some or all) become combined into a single, larger company.

The philosophy of the merger due diligence is to:

- Establish the asset values, liabilities and risks for each company;
- Determine the compatibility of the management systems used, their synergies and their differences; and
- Gather information for the development of the integration plans.

Depending on which company you represent, a merger can be thought of as a divestment (of your company assets) or an acquisition (of your partner's assets). Based on the value of the merger, levels of significance need to be established early in the process to prevent due diligence teams from concentrating on insignificant issues.

Swaps – Asset swaps are the trading of similar assets of similar value from one company to another, usually because of geographic and/or business synergies. A swap should be viewed as a divestment and an acquisition. There is often a true-up amount that is paid by one company to another if the assets prove not of the same value. Due diligence is valuable in providing accurate information required for estimating the true-up amount.

Toll Processor Agreements – are contractual agreements between a company and supplier of manufacturing services, branded products or proprietary feedstocks. A due diligence of a potential toll processor, similar to an acquisition, will provide the company information to determine the risks of the toll processing facility relative to brand values and product outage risks.

Vendor Due Diligence (VDD) – A VDD is typically conducted by the seller using a third party consultant. The VDD provides potential buyers with a brief history and current summary of the facility's health, safety, environmental and process safety programs, compliance with applicable regulations and known liabilities from the Seller's perspective.

Though other types of M&A projects exist, those not discussed above are likely to be related in way or another to those discussed above.

2.5 PROCESS SAFETY IN THE M&A PROCESS

The Process Safety (PS) professional will likely be a member of a specific M&A team, usually the Health, Safety and Environmental (HSE) team or perhaps an engineering or assets team.

The overall objectives of the HSE or engineering M&A teams are similar – capture value and manage risks through knowledge. The general strategy of the HSE/PS team is to:

- Assess and develop ways to manage the HSE/PS risks,
- Assess potential cost of HSE/PS liability and factor this into the project economics,

- Optimize allocation of the HSE/PS liabilities to maximize value and minimize retained liabilities, and
- Identify and capture HSE/PS value.

The due diligence team should:

- Raise potential issues that require special management attention to the M&A business development team,
- Provide the M&A financial team with a valuation of potential PS costs, and
- Provide the M&A legal team with sale agreement language, mechanisms and concepts that address identified PS related liabilities.

The checklist included in the Appendix of this Guideline can be used to develop a toolkit of specific concerns to be investigated as part of either organizations' due diligence process starting with the data room review and ending with post acquisition integration.

There are certain overall aspects of M&A projects that should be considered when participating in an M&A project:

- At the corporate level, what will the potential impact of the project be on:
    - The existing corporate brand, reputation and culture, and
    - Interfaces with key stakeholders (e.g. workforce, regulators, shareholders, customers, NGO's, etc,)

- From the operations perspective:
    - How the facility and company compares to internal standards,
    - What permits are in place and are they transferrable,
    - Permit flexibility and permit compliance;
    - Are there current or potential limitations to operating conditions;
    - What is the performance relative to PS, integrity management and major accident risk;

- Are there security concerns; and
- What are the capital and operations and maintenance costs, expenditure timelines and impacts?

- From the legal perspective:
  - Is the facility subject to newly enacted regulations,
  - Are there enforcement issues, permit exceptions and exclusions; and
  - Relating to the potential sales contract, what will be the representations, warranties and indemnities, as well as whether they will be shared or retained?

Prior to the close of an acquisition or potential merger, the negotiations and all activities associated with those negotiations are likely to be highly confidential. This could be the situation even where a public announcement was made that such a deal is being contemplated. If you are assigned to the due diligence process at its earliest stages, treat all information as strictly confidential even within your own organization. Should you find it necessary to seek information or advice from someone not assigned to the due diligence or M&A team, clarify with the due diligence or M&A team leader whether it is acceptable to hold such discussions. Where the team lead agrees or provides clearance to meet with someone outside these teams, you should also be clear on the extent to which information associated with the proposed deal can be shared with others.

Though some of the above aspects are not specifically related to process safety, the process safety professional needs to be aware of the overall goals of the potential acquisition or merger in order to provide feedback that is aligned with the strategy for the project to the M&A team.

2.6 FINANCIAL STRATEGISTS CAN HAVE HIGH IMPACT ON PROCESS SAFETY SYSTEMS

Financial strategists and decision makers are great in their field of expertise. However, decisions they make have rippling effects through an organization. Within the petroleum and chemical business sectors there is a basic need to have a fundamental knowledge of process safety issues. At a minimum, these decision makers need to understand there is a great deal they don't know. One major concern is - what can result from cost cutting measures?

Most incident investigators agree catastrophes occur when there is a confluence of mistakes. A prime example is the sinking of the Titanic. There has always been speculation on what could have prevented that particular disaster. The ship should have been designed and fitted with more lifeboats, the captain should have chosen a more southerly passage, etc. There are a slew of speculations on possible preventative measures that could have stopped the accident from occurring altogether or at least mitigated the consequences of striking the iceberg. However in reading the history of this epic disaster a systematic issue surfaces, namely the steamboat companies had elected to take larger and larger risks to reduce their operating costs.

Mergers and acquisitions almost always result in a strategy to cut costs. The industry needs to ensure these cost cutting measures are not directly or indirectly introducing higher risks. This basic concept is of critical importance within the chemical, petroleum and petrochemical industry sector, particularly those that handle highly hazardous substances.

Process safety awareness is critical knowledge for those situations where decisions are made for cutting costs or restructuring management after acquisitions and/or mergers. For example, in the traditional model for private equity firms they either strip costs out of their investments or bundle them with similar or complementary businesses before selling them. In addition, some private equity companies do more than acquire chemical firms. Some get involved in the company's management in order to boost the profits of their investment. Cost stripping or management decisions that disregard process safety and are made on a purely financial basis within a new organization may result in a chemical plant or refinery facility being left with limited or diluted resources. This is particularly true as it relates to understanding how to manage effectively high hazard chemical operations and address the risks associated with these hazards. When evaluating an acquisition, the financial strategists and decision makers must include appropriate process safety investment costs and operating expenses in their decision whether to proceed or not.

When a merger or acquisition involves two companies already working in the same type of high hazard industry, one would expect that a more seamless transition would occur. However, issues and concerns can also arise even with these types of mergers. The excerpts below were taken from the incident investigation report completed by the United States Chemical Safety and Hazard Investigation Board (CSB) after the 2005 BP Texas City incident.[38]

> *On March 23, 2005, the BP Texas City refinery experienced a catastrophic process accident. It was one of the most serious U.S. workplace disasters of the past two decades, resulting in 15 deaths and more than 170 injuries.*
>
> *After the Amoco merger, Texas City underwent a complex series of leadership and organizational changes that were only informally assessed for their impact on safety and health*
>
> *The management structure changed on December 31, 1998, when Amoco Corporation merged with BP. The new company, BP-Amoco, transferred corporate functions from Chicago to London. The process safety group in Chicago was disbanded and the process safety function was placed under the Technology area in the London corporate office staffed by a single advisor.*

The CSB Report of the BP Texas City incident identified a number of issues relating to organizational changes that took place following the BP/Amoco merger. The CSB Report also recommended CCPS issue management of change guidelines that address "major organizational changes including mergers, acquisitions and reorganizations." The intention of this Guideline is to help organizations work through the merger and acquisition process in such a way that process safety considerations, including possible organizational changes, are factored into decisions and allow the chemical and petroleum industry to continue to be one of the safest industry sectors.

3

SCREENING POTENTIAL CANDIDATES

If you know the enemy and know yourself you need not fear the result of a hundred battles. If you know yourself but not the enemy, for every victory gained you will also suffer a defeat. If you know neither the enemy nor yourself, you will succumb in every battle.
Sun Tzu – The Art of War

3.0 COURTNEY'S STORY - CONTINUED

Courtney entered the conference room where the kick-off meeting was being held. Most of the due diligence team was already assembled. She was mildly surprised to see "Skip", the CEO of Bland there along with Gareth the President of their European operations and Robert the President of their North American operations. After everyone grabbed coffee and a brief set of introductions all around, the due diligence team was seated and Skip proceeded with a brief introduction and overview of the proposed acquisition. Courtney didn't count but the term 'strategic importance' in one form or another had to have been used no less than ten times in Skip's remarks. Robert's remarks were more balanced. Robert noted they had learned some very hard lessons with the acquisition of Independent Refining and they did not want to 'relearn' those same lessons with the possible acquisition of White Hot Chemicals. That was why this due diligence had been expanded to include a number of new skills and experience sets, one of which was the addition of a member from Bland's human resource group and another being Courtney to examine all related HSE issues. Courtney's stomach tightened at the mention of her name. Gareth's remarks were all about the possible synergies between Bland's current UK operations and White Hot's Terneuzen operations.

After Gareth finished, Skip got up again and noted there were some legal, regulatory and technical issues that needed to be resolved before the full diligence process could begin. They were hoping to have those matters finalized not later than the end of the following week. In the meantime he wanted the due diligence team to gather as much information as they could on White Hot. He was very interested to see if there was anything 'really big' out there Bland should know about as the negotiations continued. Skip told them use whatever sources of publicly available information you can – but you can't contact anybody in White Hot, at least not yet.

3.1 USING PUBLIC DOMAIN INFORMATION FOR SCREENING

You are sitting in your office at 3:00 pm on a Thursday afternoon and the phone rings. The caller tells you that you are on the team for an acquisition of White Hot Chemical Company's plant in Houston. What do you do? Do the same thing that you would if you wanted to know something about a Labrador retriever – Search the Internet! So you type in: White Hot Chemicals, hit 'enter' or 'search' and a world of information opens to you. Congratulations you have started your due diligence.

There are sources of information on the Internet that can jumpstart a due diligence process. These include:

- The target company's website;
- An internet or Web search using the company's name;
- Federal, state and local regulatory agency websites;

- Newspaper and television media sites local to the plant, and
- Maps and aerial or satellite photos from sites like Google Earth® or Microsoft Virtual Earth®.

In a couple of hours a significant amount of background information can be collected on a company, as well as individual plant and plant sites starting with only the company name and site location. *It is important to note however,* that *this background information* should be confirmed either during the site visits or from materials and documents obtained through the data room. Similarly, news articles might not be completely factual and may include opinions or commentary. Use them as red flags or for guidance about an issue, but not necessarily as facts to be included in the final due diligence report.

A good place to start is the target company's own website. Here, if lucky, you might find information on the safety culture of the company that will give you insights into what you might find when doing your plant site visits. Many companies have their HSE policies, and in some cases details about their management system on their sites. Where this is the case, these should be reviewed to see if process safety is included. Look for information regarding process safety metrics. CCPS recently issued a publication on process safety metrics.[39] Some key leading metrics to look for include:

- Mechanical integrity tests and inspections completed
- Percent of time plants operated with safety critical equipment out of service
- Action item follow-up to final close-out
- Management of Change properly implemented
- Percent of startups following changes that encountered process safety problems
- Process safety training completed for critical PSM positions
- Training competency assessments

- Percent of safety critical tasks where procedures were not followed
- Safety culture assessments

Also do a search within the company's website(s) for these lagging process safety metrics:

- Process Safety Total Incident Rate (PSTIR)
- Process Safety Incident Severity Rate (PSISR)
- Near Miss Reporting Information System implemented (NMRIS)

You might even be able to search the website for the plant in question, and find facts concerning products made and other key information. A company's annual report and sustainability report, if they prepare one, might have sections on safety and process safety and these reports are usually posted on a company's site, especially on the websites of publicly traded companies. While reviewing the annual report, make sure to review the 'footnote' sections to the financials as these will often list significant regulatory action(s) or pending litigation that may have arisen as the result of an incident or accident. If in a search of the company's website you find the street address of the plant, take notes as it will help in further searches.

If the plant has had recent accidents or fires, an Internet search will likely have various news links listed. News stories will provide an idea of the plant history, and the tone of the articles will help in identifying the relationship between the plant and the local media. News articles will also indicate whether these accidents led to regulatory sanctions, fines, litigation, etc. Note the dates of any fires, releases of chemicals or materials to the environment, notable emergency shutdowns, etc. within the past three to five years. Add these events to the list of information you as a potential buyer want provided in the data room. Then review the associated incident reports developed by the company as part of your due diligence. Incidents outlined below serve as red flags and provide an idea of the status or effectiveness of the process safety program of the site:

- Major equipment malfunction(s),
- Significant corrosion, and
- Large diameter line, tank or vessel ruptures.

Do not limit your initial search to the first screen of links. Check through the pages of links until it is obvious you are no longer finding useful information. This will help to develop a history of the site. If the initial search does not list any news articles, conduct a search on something like "news media (city of choice)" – then search the individual news sites that are returned using the company or a plant site name.

Another good Internet site for process safety type information, specifically major accident risk, is one that has aerial photographs such as Google Earth® or Microsoft Virtual Earth®. With a street address you can find almost any structure in North America and Europe, plus there are satellite photos of almost everywhere on Earth. Once you find the plant site, you can see unit spacing and what surrounds the plant from a major accident risk perspective. If the plant or facility is in an area with a satellite photograph you can 'zoom-in' on the site and obtain a perspective on:

- The number, placement and general size of storage tanks,
- Rail and truck loading racks,
- The general size of process trains and larger process vessels,
- Onsite equipment layout,
- Location of both occupied and portable buildings,
- Watercourses and adjacent facilities,
- Other neighboring community issues.

This may provide insights into facility siting issues that will need to be confirmed later in the due diligence process both by reviewing any facility siting analyses carried out by the target company as well as when conducting actual site visits. You can also make some extrapolations about future land use in the area. If the plant is surrounded by farmland but a higher view shows new housing developments in the area, you might consider the potential consequences on the operations of the site if new housing was built next to the plant. This might mean land immediately adjacent to the site will have to be purchased within a few years as a buffer. A word of caution, some countries consider aerial or satellite photos of chemical plants as espionage – do not email aerial photos to the other due diligence team members, as some of them might be in such a country.

Federal, state and local agency Internet sites are also a good source of information. These give information on permits, but also many sites contain permit applications and environmental impact assessments for a number of projects that can be viewed for background information and ideally, process safety information. These sites are sometimes more difficult to search through, but useful information may be found. Some states have their own regulations with process safety requirements that go beyond those required by OSHA and EPA. For example, New Jersey under their Toxic Catastrophe Prevention Act Program has requirements for reactive chemicals management and inherently safer technology analyses. Some regulatory web sites within the U.S. you should check include:

Table 1. Several US based Agencies and their Website addresses

Agency	Website Address
US Environmental Protection Agency Office of Emergency Management	www.epa.gov/emergencies
US Occupational Safety and Health Administration	www.osha.gov
US Department of Transportation	www.dot.gov
US Chemical Safety and Hazard Investigation Board	www.chemsafety.gov
US Department of Labor	www.dol.gov
New Jersey – Toxic Catastrophe Prevention Act Program	www.nj.gov/dep/rpp/brp

Other states with significant additional process safety regulations include California, Nevada and Delaware. The websites of these and other state regulatory agencies should be researched as well.

Where a potential acquisition includes sites in other countries or regions, check the websites of the respective regulatory agencies in those countries. The following table of such agencies is by no means exhaustive but more to give you an idea of websites operated by regulatory agencies in other countries or regions.

Table 2. Examples of International Agencies whose Websites should be checked for Process Safety Information

State or Region	Regulatory Agency	Website
United Kingdom	Health and Safety Executive	www.hse.gov.uk
	Health and Safety Executive prosecutions	www.hse.gov.uk/prosecutions
	Environment Agency	www.environment-agency.gov.uk
European Union	European Environment Agency	www.eea.europa.eu,
	Seveso Inspections	http://139.191.1.51/typo3/index.php?id=78
	European Agency for Safety and Health	www.osha.europa.eu/en
The Netherlands	Ministry of Housing, Spatial Planning and the Environment	www.vrom.nl
	Staatstoezicht op de Mijnen	www.sodm.nl/English

(continued)

Table 2. Examples of International Agencies whose Websites should be checked for Process Safety Information

State or Region	Regulatory Agency	Website
Canada	Canadian Centre for Occupational Health and Safety	www.ccohs.ca/products/databases/fatality reports.html
Malaysia	The Department of Occupational Safety & Health	www.dosh.gov.my
India	The Directorate of General Mines Safety	www.dgms.gov.in,
	The National Safety Council of India	www.nsc.org.in
Australia – Commonwealth Level	Safe Work Australia	www.safeworkaustralia.gov.au
	National Offshore Petroleum Safety Authority	www.nopsa.gov.au
Australia – State Level	West Australia Department of Mines and Petroleum	www.dmp.wa.gov.au
	WorkSafe Victoria	www.worksafe.vic.gov.au
	New South Wales - WorkCover	www.workcover.nsw.gov.au

As you can see from this table, you might have to do a little digging to find all the regulatory agencies in a particular country or region that have administrative responsibility for process safety or major hazard related issues. Further, process safety issues as commonly referred to in the United States are typically called or referred to as 'Major Hazard' issues in Europe and the UK Commonwealth countries. When searching for legislative requirements be aware that PS issues may well be regulated under dangerous goods or dangerous substances regulations while in other cases they may be regulated under emergency management laws or regulations. When conducting searches of sites in other countries be sure to use these terms as well.

Another screening criterion is whether the company is involved in various industry organizations that promote process safety. One such site is the American Institute of Chemical Engineers (AIChE) and the Center for Chemical Process Safety (CCPS). Others such organizations within the U.S.A. include:

Table 3. Associations that promote good Process Safety practices

Association	Website Address
American Chemistry Council	www.americanchemistry.com
American Institute of Chemical Engineers	www.aiche.org
American Petroleum Institute	www.api.org
American Society of Mechanical Engineers	www.asme.org
Mary Kay O'Connor Process Safety Institute	www.process-safety.tamu.edu
National Fire Protection Association	www.nfpa.org
National Petroleum Refiners Association	www.npradc.org

Companies who are active members in these organizations will generally have better process safety programs than companies who do not participate. A more detailed search may include representation on various process safety committees by company personnel. A company where individuals actively participate on committees and present papers at conferences is more likely to have higher quality process safety programs in place. Active participation on technical committees, conferences, presenting papers, etc. by a company's staff shows the management of the company recognize the value of such participation and are committed to supporting best industry practices in their own operations.

Further check the websites of various Non-Governmental Organizations (NGO's) such as Greenpeace, the World Wildlife Foundation, National Resources Defence Council, Stockholm Environment Institute, etc. Similar to the discussion of using news articles to identify an issue you will want to examine further, information gained from these sites also needs to be thoroughly investigated as the due diligence process proceeds before including in any reports.

Table 4. Examples of Non-Governmental Organizations and their Website addresses

Non-Governmental Organization	Website Address
Greenpeace	www.greenpeace.org
National Resources Defence Council	www.nrdc.org
World Wildlife Foundation	www.worldwildlife.org
Stockholm Environment Institute	www.sei.se

When conducting these searches, if you find a major problem inform the due diligence or M&A team leader(s) immediately. For example, a recent news article about a local town council voting to shut down the target facility should be addressed quickly. Other red flags may show up in the preliminary searches, which the due diligence lead(s) should factor into their decisions to continue with the acquisition or merger. M&A projects are people resource intensive and can be expensive. The M&A team leader will want to shut down bad prospects early in the process where necessary. Some process safety issues that may derail an M&A project at this stage might include:

- Significant PS issues that were identified as part of a regulatory agency inspection such as a recent US-OSHA, National Emphasis Program (NEP) inspection or a Risk Management Plan inspection by the US-EPA
- Significant facility siting issues requiring major building relocation or modification,
- An ageing plant or equipment with poor maintenance history,
- History of incidents, especially those impacting offsite or neighboring facilities, or
- Proximity to sensitive populations such as schools, hospitals, recreational areas, or sites deemed to be of special environmental significance.

If you are involved with a divestment, conduct an Internet search on the facility you are divesting. If there are news articles about the plant site in a Google search, ensure proper information on such situations is included in the data room and the Vendor Due Diligence report. Remember if you found it via an Internet search, a savvy potential buyer or bidder will have found it as well. As a result any relevant information will be asked for in a formal Q&A bidder response or as part of site visits. There may be a significant amount of information on a company's internal website relative to a particular site, facility or business unit. This information should be obtained in a confidential manner and used to populate a data room, as well.

For a Joint Venture or merger, an Internet search should be conducted on the partner organization and their assets. You may find the HSE Policy and management systems on the partner company's website. There will likely be other indicators of their standards and corporate direction. One added step is to search the key company members of the target company's board of directors. Board members are generally listed in the annual report. You may find an individual director is involved with a process safety issue that you might not want to assume the liability for.

Other web services that are usually paid subscription services may be available to your law department or the legal members of an M&A team. They should be searched to gather information on any pending lawsuits, enforcement actions and settlements within the past three to five years. While these searches might produce good information, at this screening stage and level of review it might be best to ask the legal team or your own law department if they have run any such searches and whether those searches have identified any potential 'deal killers'.

3.2 USING A CHECKLIST TO IDENTIFY POTENTIAL PROCESS SAFETY ISSUES

Section 3.1 provides guidance on various means you might use to collect publicly available information as part of screening potential acquisition or merger candidates. A checklist of process safety issues has been developed to assist with the type(s) of information that should be collected as the acquisition and/or merger moves forward. That checklist is included in Appendix A to this guideline. The checklist is organized based on the various phases of the M&A process including:

- Commercial Phase
- Establishing the process safety expertise on the M&A Team
- Information to be included in a Data Room
- Information that should be collected as part of planning the site visits, and
- Information or issues to be examined as part of the site visits

It also includes guidance on specific process safety issues in the following areas:

- General Process Safety
- Assessing Major Risk
- PS Management Systems & Culture
- PS Staffing
- Hazard Identification
- Managing Changes
- Process Safety Information
- Mechanical Integrity
- Procedures
- Audits

The checklist covers approximately thirty-eight pages (landscape) of, for the purpose of this guideline, process safety related 'pointers'. 'Pointers' an individual(s) should consider when undertaking an acquisition or merger of a business or facility that stores, processes, transports, etc. materials that are highly hazardous. However, despite the fact the checklist is lengthy, it is not intended to be exhaustive nor to cover all process safety related issues or situations entailing an acquisition. There will be unique factors in every acquisition or merger that cannot be readily covered in a generic checklist. Nor should the checklist be used assuming every item in it has to be addressed in every acquisition or merger. It is to be used intelligently, by an individual or individuals who have experience and knowledge in the process safety area. By melding the experience of those individuals with the 'pointers' provided in the checklist, it is hoped they can then develop a checklist of issues that is fit for the purpose of the acquisition or merger that lies in front of them.

4

THE DUE DILIGENCE PHASE

Only when the tide goes out do you discover who has been swimming naked.
Warren Buffet

4.0 COURTNEY'S STORY - CONTINUED

This was the third time the due diligence team was meeting. The first was the kick-off meeting, the second about five days later after the due diligence team had a chance to do some preliminary investigations on White Hot and presented what they found to Robert and Gareth. Courtney brought up the OSHA PSM violations along with some concerns about the physical locations of the Terneuzen, Netherlands and Pasadena, Texas plants. She had also found various articles in local news reports and newspapers on releases that occurred from the plants in Pennsylvania and New Jersey. There didn't appear to be any investigations of these, at least as far as she could determine from any of the regulatory agency websites. Robert noted he wanted to make sure these points were followed up on during the actual due diligence investigations which had now been agreed. However, Robert went on to note they didn't sound like they were potential 'deal killers'. Robert did continue that all the same, they had some real nasty surprises surface after they had closed the deal on Independent and they did not want to repeat those mistakes.

Josh, one of the directors in the business development group, was leading the due diligence team and this third meeting. The focus of this meeting was to plan out the actual due diligence investigations. Josh noted the timeline set out for the due diligence was tight. They were well into the third quarter of the year and Skip the CEO of Bland was anxious to complete the deal by the end of the year. The first thing Josh wanted from each of the team members was their critical issues along with a list of information, materials or documents related to those issues they should request from White Hot. He was especially anxious to get that list from the HR representative and Courtney as this was the first time HR and process safety issues were being addressed in the due diligence investigations. Josh turned to them both and asked, "Would tomorrow be too soon?"

4.1 INTRODUCTION

The costs to rectify Process Safety concerns or issues identified after a deal closed and integration of the assets began in earnest, has been reported to have run between ten and thirty percent of the initial purchase price. We hope by the time you finish this chapter let alone the Guideline as a whole, you will not be one of those individuals caught out by an ebbing tide as observed in the opening quote of this chapter. Caught-out because of holes in your due diligence process.

In Chapter 2 the flow of the M&A process was introduced and Chapter 3 discussed using the Internet to begin collecting information as part of the due diligence process. In reality, however, the due diligence starts with developing a checklist of issues that will need to be investigated as the due diligence process moves forward.

To assist, a generic Checklist of M&A Process Safety related issues has been prepared and included as Appendix A of this Guideline.

When working on a divestment or an acquisition, the use of such checklists will be the same but the order of events is different. The major difference is that in a divestment the site visit(s) and document reviews are conducted to develop a Vendor Due Diligence (VDD) report and populate the Data Room. In an acquisition, the Seller completes the VDD and Data Room first. The Buyers after reviewing information in the VDD and data room follow this up with site visits and the preparation of their own acquisition due diligence report (see Figure 1).

Both the divestment and acquisition processes start with developing a checklist. Checklists should be living documents that are modified in light of changes in regulations, standards or industry practice as well as where an issue or set of issues gets more complex.

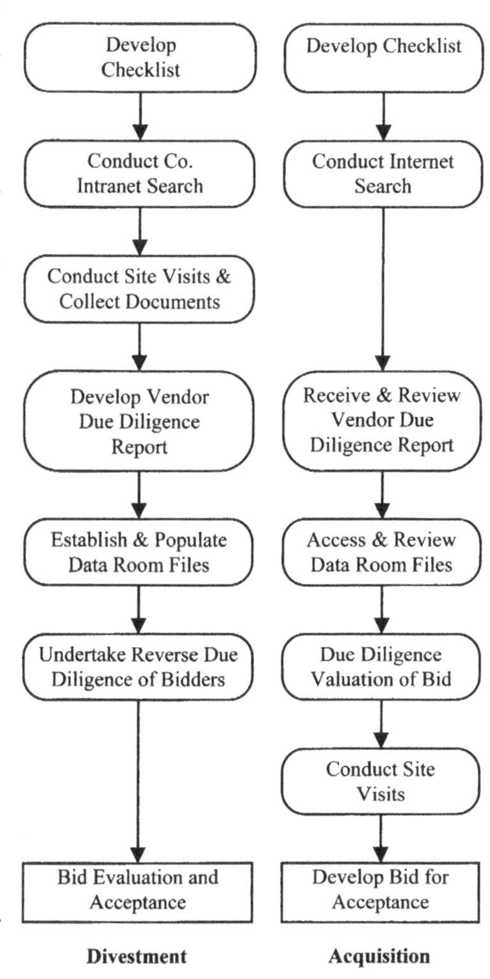

Figure 1. A comparison of steps between a Divestment and Acquisition

Further, you should view every potential acquisition whether the deal closes or not as a learning exercise for your company. Lessons learned in the due diligence phase should be captured and included in a checklist you have developed that is unique to your particular company.

The objective of any due diligence exercise is to identify issues, evaluate them and then quantify them in some manner, usually in the form of liabilities, and the potential monetary impact those liabilities could have on the value of a particular transaction. Due to the potential magnitude of most acquisitions or mergers and the impact of on-going liabilities could have following the close of these deals, third parties are often brought in to assist with the due diligence process. In the HSE area, this is often as a result of concerns regarding potential environmental legacy issues surrounding or associated with the operations of the target assets. Since potential process safety issues could also significantly impact the potential value(s) of an individual site or group of assets, you might want to consider retaining an independent third party to assist with identifying and evaluating PS related issues as well. If your company does retain a third party, you should check to assure yourself that third party has not been involved in any activities relating to the site(s) or company being acquired or divested. Also remember that a due diligence is <u>not</u> just another form of an audit. While the development of a protocol or checklist of issues is advocated in this Guideline it is there to supplement and establish a level of consistency when examining a variety of issues. If the only output a due diligence team provided to the M&A project lead or M&A negotiating team was a series of completed audit protocols, that would not assist those teams with the tasks they have to accomplish. Of critical importance in staffing the PS part of the overall due diligence team is the ability to establish what, when and where an issue or group of issues could materially affect a proposed deal. The ability and experience to identify and quantify material deficiencies in a process safety program, is a key skill to be considered whether resourcing the due diligence team using internal staff or using an external third party.

4.2 THE DIVESTMENT DUE DILIGENCE

The philosophy of a divestment due diligence is to provide potential buyers with factual information about the assets in a timely manner. This is accomplished by preparing a factual summary report, the Vendor Due Diligence (VDD) report, as well as populating a data room with information that provides a potential Buyer with an understanding of key issues associated with the business, operation(s), individual site or facility.

The VDD provides potential buyers with a brief history and current summary of the facility's health, safety, environmental, security and process safety programs, compliance and known liabilities from the Seller's perspective. The VDD goes beyond a traditional Phase I Environmental Assessment, which typically focus on soil and groundwater issues. The VDD should provide an overview of the general health, safety, security and process safety aspects of the facility. It should include positive aspects of the process safety and the broader existing HSE programs and accomplishments for the facility. It should also disclose known issues and liabilities that may require on-going or future expenditures. The VDD provides information to the potential Buyer that can be used to establish the asset values prior to preparing their bid and their full due diligence.

The VDD also provides the Seller's business team with a summary of the known HSE issues to consider for inclusion in the marketing and negotiating plans as well as potential costs and liabilities that may impact the sales price of the assets. During the VDD it is helpful to establish a level of significance or "materiality" (e.g. $100,000, $1 Million, $10 Million) so the due diligence teams can focus on "material" issues above that level.

Existing liabilities can be handled in a number of different ways relative to the divestment. The way in which they are handled is very important to the contract negotiations and the due diligence process. They should therefore be established early in the project. If the Buyer is to assume the liabilities, the VDD must define and discuss the known liabilities and provide sufficient information for the potential buyer to estimate the cost of each. An alternate method is for the Seller to retain the pre-closing liabilities, in which case the VDD should concentrate on defining the existing liabilities and establishing a baseline cost.

4.2.1 The Checklist

A checklist of all reasonably foreseeable issues is the key to organizing a successful due diligence. The checklist helps with ensuring key issues are not overlooked during the due diligence. With some knowledge of the facilities in question and expertise in process safety the due diligence team member can use the checklist to help guide them through the divestment process. The checklist in Appendix A can also be used as a guide for the Table of Contents of the VDD report and the Data Room index where your company does not already have a standard format. Where the due diligence activities are to be conducted at multiple sites or by using multiple teams, a checklist will help the VDD reports to be consistent with one another. It will also help in ensuring a level of consistency when collecting information from the various sites for inclusion in the data room. However, checklists should not be solely relied on to identify all potential process safety related issues that could be associated with a potential merger or acquisition.

4.2.2 The Internet and Intranet Searches

Conducting an Internet search on the assets being divested as discussed in Chapter 3 can reveal red flag issues, such as, negative newspaper articles regarding facility accidents or incidents that should be factually addressed in the VDD and/or the data room. It is quite likely these issues will be raised in the Q&A portion of the data room by the Buyer and it is usually easier to address them while the resources of the due diligence team are fully functional rather than wait until after they have been disbanded. These specific issues can be added to the checklist to ensure they are addressed. It is also beneficial to search the Seller company's external website to determine if certain documents are available, such as, the company's policy statements and HSE management system. Often during the Data Room Q&A or site visits the Buyer will request copies of these documents. Rather than going through the legal team to have these documents copied, it is much easier to give the Buyer the link to the company website to review the documents.

A review of the company's Intranet site may produce many documents related to the assets that can be downloaded, reviewed for questions, provided in the data room and stored prior to any site visits. As site visits are time intensive, a review of such documents prior to the visit can expedite the process. Downloading documents from the company's Intranet eliminates the need to obtain copies at the facility itself during the site visit(s). This will lessen the strain on the local staff and the due diligence resources, in general. Items that require specific questions to be addressed during the site visit can be added to the checklist. Other key information, such as a list of the facility safety procedures and copies of significant process safety procedures, can often be downloaded from the facility's intranet site, as well.

The divestment of a facility or group of assets within a company is usually extremely confidential at this stage of the M&A process. As a result, contact with facility staff should not be done unless approved by the M&A Project Manager. Intranet searches can be conducted while maintaining that confidentiality.

4.2.3 Pre-site Visit Review

It is possible that in the preparation of the Seller's VDD, it is decided that various sites will be visited as part of the development of the VDD. The Seller may also agree to a series of site visits by the Buyer to assist with the Buyer's preparation of a DD report. The process for conducting both visits will be similar.

Prior to the site visit, the Process Safety reviewer should compile a list of individuals or functional staff for interviews and submit the list to the due diligence project manager. Also prior to the visit, the Process Safety reviewer should be aware of the personal protective equipment (PPE) and other safety requirements required by the facility's safety policy. The reviewer(s) should, as far as practicable, arrive at the site with all required PPE ready to perform the assessment. Most facilities will require a more detailed orientation for unescorted versus escorted visitors. Time for orientation should be factored into the schedule of the site visit(s).

Larger facilities with their own support staff will often maintain records for the facility on site. These can be reviewed and copied/scanned/downloaded for the data room during the site visit. Records for smaller facilities might be kept in corporate offices or other locations remote from the facility. Prior to the site visit, the location of required information should be determined. If the records are not at the site, it is preferable they are located and reviewed prior to the site visit.

This guideline will not go into the details of conducting audits of the records or the site visits of which a due diligence is one type. For assistance on conducting such audits, refer to the CCPS auditing guidelines book.[40]

4.2.4 The Due Diligence Site Visit and Document Review

The site visit is often considered to be the key aspect of the due diligence process. The purpose of this visit is to address, update and validate all the questions and assumptions made by the PS reviewer prior to that time.

The site visit for a large facility can take from three to five days. At a small facility with few support staff the visit may only require a day or two. Table 1 outlines the activities typically undertaken as part of a site visit:

Table 1. Typical set of activities undertaken as part of the site visit

Typical Site Visit Agenda
• Safety orientation • General introductions of the Due Diligence (DD) team and site staff members • General facility overview, plot plan and locations of key staff members • Logistical requests from the DD team, such as, reiterate the list of key staff for interviews and ask their availability, the DD team schedule, location of the facility files that are needed for review and check-out procedures, options for lunch, etc. • A quick driving tour of the facility so the DD team can get the "lay of the land". • Initial interviews; an interview with the person responsible for the overall Process Safety program should be scheduled early in the visit to get an idea of the scope of the program and the names of other individuals who should be interviewed as part of the visit. • Two to three days of interviews, data and document reviews and individual facility inspections. Refer to the checklist often during this process to ensure you are capturing all the requirements and issues. • After hours, prepare sections of the DD report while the information is fresh in your mind. • Closeout or exit meeting with the site staff at the end of the DD site visit. • Initial closeout meeting between the members of the due diligence team where the team members are unlikely to have the opportunity to meet again face-to-face.

When discussing and gathering reports, and other materials for the data room, documents should be collected that cover all aspects of issues. For example, if a hazard assessment of one of the operating units was conducted, the hazard assessment report, the action plan, and the current status report for the action items should be included. If similar assessments have been conducted on an individual operating unit over the years, only the most recent assessment report, action plan and action plan status report is normally included in the data room. The existence of previous assessments can briefly be mentioned in the VDD report. If a report, document or spreadsheet is being discussed during an interview, the reviewer should obtain a copy at that time for further review and possible inclusion in the data room, where possible. If the document is stored electronically, it is often convenient for the reviewer to copy the document to a portable data storage device, e.g. a CD or flash drive. Where documents are provided electronically they should be provided in a Portable Document Format (e.g. as a .pdf file) as opposed to a Microsoft Word® or Excel® spreadsheet to help with preserving the integrity of the original document.

Reports and other documents should be reviewed during the site visit to ensure that questions the reviewer may have regarding the report or the findings are addressed during the site visit interviews. In a divestment, the reviewer's duties are not to conduct a formal hazard assessment of a facility. Rather the reviewer's objectives are to report the existence of any such assessments previously conducted by others and the status of the action items.

If the reviewer is concerned about compliance with a regulatory provision, this concern should be discussed with the facility staff to obtain their explanation prior to documenting whether the site is in compliance or not. Often the reviewer will not be aware of nuances or unique situations in a facility's compliance plan. Where they are identified they should be explained in the VDD. All non-compliance items discovered by the reviewer should be communicated to the M&A business team immediately to allow for possible correction of the non-compliance issue prior to the marketing/sale of the facility.

One concept often overlooked in the DD activities is the documentation of the awards and meritorious recognitions a facility has received from third party organizations. These awards show a facility's performance has been recognized by others and may add value to the sale. They are often displayed on plaques on the wall in the foyer of the administration building, in training rooms or in the HSE department offices. Take note of such awards and mention them in the interviews to get background information that should be included in the VDD report and data room. If possible, obtain documentation relating to the awards and/or readable pictures of the plaques. It is often acceptable to take the plaque off the wall and make a photocopy of it for scanning if a picture cannot be obtained. These awards are positive additions to the data room.

4.2.5 Vendor Due Diligence Report

The VDD report is a factual report. The reviewer should report the facts and compile relevant documents for the data room.

> *Example: Using our previous example of a hazard assessment of an operating unit, the text for the report could look something like: "A hazard and operability (HAZOP) study was conducted for the catalytic cracker in 2008 by members of the site engineering team with assistance from ABC Engineering. The HAZOP generated 87 action items. At this time, 50 of the items have been completed, 20 items have been engineered and added to the routine maintenance list, and 17 have been added to the work list for the scheduled unit shut down in 2010. Costs for these items are included in the 2009 maintenance budget and the 2010 turn-around of T&I budget." Existing and forward budget data is usually included in the financial section of the data room and need not be documented in the VDD report.*

The VDD report needs to address all material issues or concerns that were to be investigated per the checklist originally developed. Unless the company has a specific format they wish to use for the Report, the VDD can be organized using the same general structure as the checklist. Organizing the data room in the same order as the VDD report will help when cross-referencing sections of the Report to the data room documents, as well as during the data room Q&A process.

In writing the Report, it should be recognized those who will be reading it may have their own checklist(s). For this reason, as much information as possible should be communicated in the Report, even for issues that might seem to be tangential or of little significance. For example, if the facility is not covered by PSM or Major Hazard regulations, the Report should state something like:

> *"The facility is not covered by the PSM regulations because the quantities of covered chemicals handled onsite do not exceed the specified threshold quantities."*

The checklist also notes that the reviewer should request a list of major equipment with age and general condition with reference to maintenance schedules. If no such master list exists, the VDD might include a statement similar to the following:

> *"Equipment lists for each operating unit are maintained in the maintenance department per functional group, such as Pumps and Drivers, Pressure Vessels, Tanks, Heat Exchangers, etc."*

Though these points may seem trivial, during the data room Q&A period the questions from this checklist will likely be asked and it is much easier to answer the questions by referencing the existing answers in the VDD report.

The VDD report is a technical report and should factually represent the positive and negative aspects of a facility or the whole business being sold or bought. Further, the overall nature of the Report should highlight what the facility has done, rather than stressing what is lacking. For example, referring back to the hazard assessment example it would be factual to state – "The HAZOP generated 87 action items of which 37 remain open". However the more positive statement used in the example is also factual. Other positive phrases can include: "45 percent complete" rather than "55 percent incomplete", or "75% of the staff have received training" rather than "25% of the staff have not received training".

Once the VDD report has been drafted it should undergo a technical review by one or two key members of the facility staff to ensure it is accurate. Key staff members might include the site HSE manager and the engineering manager. As this is a key document for the data room, it must be technically accurate and capture all material issues that are required to be disclosed to the buyer.

The VDD provides the Seller's business team with a summary of issues for the development of their marketing and negotiating plans and a background for assessing bids. As a result, the PS specialist will need to include their views, observations or findings concerning the state of the current systems and physical equipment and how they compare with regulatory requirements and good industry practice (i.e. how the assets and systems compare to Recognized and Generally Accepted Good Engineering Practice {RAGAGEP}). Gaps in this regard need to be highlighted to the seller's M&A business team to assist with their reviews of a potential buyer's bid submission. As an example, where the Buyer will be obtaining insurance for the facility, the VDD can be a key document in the underwriting process that can help accelerate with closing the deal. The VDD report will also become the historical document related to the sale. This can be extremely useful in defining what issues were disclosed prior to the sale if legal actions are taken after closing.

4.2.6 Valuation

The prospective buyers will submit bids and general contract terms to the Seller's business team as part of the selling process. To assess these bids, the Seller's business team will need to know the liabilities associated with the operation and the future spending requirements for those liabilities to determine the present value of the business.

The most significant costs in bringing a facility into compliance with the process safety regulations including all applicable industry standards are generally those related to repair, replacement or modification of the physical plant, equipment, buildings, control systems, etc. Issues or items that could significantly impact the cost of a potential M&A, the PS reviewer should examine include:

- The actual physical state of the fixed equipment (e.g. tanks, vessels, reactors, etc.) and associated piping and the standards by which this equipment was inspected, repaired and maintained,
- The actual physical state of all prime movers (e.g. large pumps and compressors) and the standards to which this equipment has been inspected, tested and maintained,
- The location of all occupied buildings to the processes and or storage facilities including offices, control rooms and workshops and whether these buildings meet current industry standards,
- The overall design of and the actual state of the relief system(s) and whether these are in accordance with current standards,
- Whether the Safety Instrumented Systems meet the requirements of current standards.

These are all areas where the investments required to repair, replace or upgrade these items to comply with current standards could accrue to a level approaching or possibly exceeding ten percent of the estimated purchase price of the proposed transaction. To determine the potential liabilities, the PS reviewer should first request the status on these points during interviews with the site engineering staff. Where the need for repair, replacement or upgrade has already been determined information on the cost estimates and timing for completion should be requested. Where, arising out of the PS reviewers own judgment issues exist and the site or plant has not made provision for addressing them, the PS reviewer will either need to develop their own estimate or obtain permission to meet with their own company's engineering staff to develop those estimates. These valuations will then be forwarded to the business team to be included in their financial evaluations. The valuations should note whether or not the process safety projects are included in existing capital or operating improvement budgets the of the target site or business or whether additional provisions must be made to complete these projects. **This detailed summary information is extremely confidential and should only be shared within and among the business team. The summary should not be included in the VDD report or data room.**

4.2.7 Data Room

In a divestment, the first level of information initially provided to prospective buyers is usually a brief information memorandum used to market the business. This document may or may not contain information relating to process safety. The second level of information is the VDD report. The third level is the data room. The fourth level of information the Buyer will gather occurs during the site visit(s) where almost all documents at the site will be available for their review. As such, the data room should contain the supporting documents and information mentioned in the VDD report. If a document is mentioned in the VDD report and is not in the data room, it is likely to be requested during the Q&A period or during the site visit.

The data room can be a physical location or room, or an electronic repository of all the documents collected during site visits and pre-site visit(s) activities that will be compiled by the seller and shared with the potential buyers. The materials in the data room should support the first two levels of information as noted earlier made available to potential buyers.

Check with the business team's legal representative or your legal department to determine what types of documents are planned to be included in the data room, how the data room will be initially populated with documents and how it will be managed in the future. The type of process safety information to be included in the data room should also be agreed early with the business team lead.

The types of documents usually included in a data room relating to Process Safety are summarized below:

Table 2. Outline of Process Safety Related documents typically provided in the Data Room

Documents Typically Provided in a Data Room

• The Vendor Due Diligence report,
• Safety awards and recognition for safe operation from third parties,
• Periodic HSE reports if they address process safety,
  o These may be in another section of the data room (e.g. under technical or mechanical integrity) in which case they need not be duplicated in the PS section.
• Site plot plan(s) (or generally arrangement drawings),
  o If engineering drawings or piping and instrument diagrams and other drawings are to be included, they should be included in the engineering section of the data room rather than the process safety section. The same holds true for equipment lists and specific process information.
• Management systems relating to Process Safety
• Process Safety metrics and status
• Process Safety hazard assessment reports, action plans and status reports.
• Risk assessments, studies, action plans and status reports
• Reports from the insurance carrier's risk or engineering departments,
• Audit reports, action plans and status reports
• Incident reports, Root Cause Analysis reports, action plans

Documents Typically Provided in a Data Room
and status reports • Permits and compliance reports • Correspondence with regulators and governmental agencies • Notices of Violations, Court Orders, Administrative Orders • Listings of safety, operating, and emergency procedures used at the site. o For the data room, this may be a copy of the Table of Contents of the procedures manuals or computer based system. • Key policies and procedures that relate to process safety, such as, Management of Change, Process Hazards Analysis procedures, etc. • Lists of training programs and requirements. o These may be included in the procedures manual(s) or in the training department. • Equipment testing and inspection program overview documents. o If this does not exist, an inspection program discussion in the VDD may be an acceptable alternative.

The documents in the data room are pre-existing documents available from company or site records. Except for the VDD report and some of the financial information in other sections of the data room, documents are not normally generated specifically for the data room. If a document listed on the checklist does not exist, such as a master list of all the equipment on site, it is not necessary to generate it just for the data room.

All documents in an electronic data room should be in a Portable Document Format (PDF) or similar format. Where a large spreadsheet is to be provided, it should be in a Read Only format. In preparing the electronic documents for the data room it is very useful to develop a consistent nomenclature for naming the files. This nomenclature will allow the buyers' reviewers to more easily maneuver through the data room as they will not need to open each file to determine its contents. It will also be easier to find particular files when responding to any Q&As. For example the nomenclature taxonomy could be structured along the following lines – *Operating Unit _ Study_ Date.* Referring back to the hazard assessment example, the file names would be: "Cracking Unit_HAZOP_06062008"; "Cracking Unit_HAZOP Action Plan_06152008"; and "Cracking Unit_HAZOP Action Plan Status_01152009".

The data room is the repository for all documents to be shared with the buyer(s). As such, additional documents will be added as each of the Buyer's DD progresses. Depending on the nature of the data room, the new documents can either be added into the index framework of the initial data room or added by creating a new documents file. The addition of new documents should be considered when establishing the initial data room index for process safety documents.

At the conclusion of the divestment, the data room documents will be archived. It is extremely important that all documents supplied to the buyer prior to closing are in the data room so when archived the document lists are complete.

4.2.8 Question and Answer Management

The potential buyers reviewing the data room information usually have an opportunity to ask questions or request additional information relating to the business via a formal, written Q&A process. The questions will likely be submitted to the Seller's negotiating team lead, who will then farm the questions out to either the lead of the Seller's due diligence team or directly to the Seller's due diligence reviewers. Generally, the answers to the questions will need to be developed in a very short time frame. As the questions and responses will be included in the data room, they too must be technically accurate. As a result, it may be necessary to contact individual facility staff to assist in answering the question(s). In many cases the answer to a question may be in the VDD report or in your notes on the checklist.

The first assumption to make when responding to questions is that the person asking the question has not been provided a copy of or has not read the VDD report or possibly overlooked the section on Process Safety. This could easily be the case if the VDD report is entitled something like HSE Assessment and the process safety reviewer for the buyer did not think it applied to their area of expertise. If you find the question is answered sufficiently in the VDD report, your answer should direct the reviewer to it with section and page number. You might cut and paste the text from the VDD report into your answer but avoid paraphrasing the language in the VDD as this becomes "new" information. For example, per the previous VDD example, a buyer's representative may request "a list of major equipment items including age and general condition and references to preventive or condition-based maintenance schedules." If no such master list exists, and you have noted in the VDD report; "Equipment lists for each operating unit are maintained in the maintenance department per functional group, such as Pumps and Drivers, Pressure Vessels, Tanks, Heat Exchangers, etc.". Then you might state "As noted in the (VDD report name and date) located in the data room in Section XX, in report section 6.2 page 32 – "Equipment lists for each operating unit are maintained in the maintenance department per functional group, such as Pumps and Drivers, Pressure Vessels, Tanks, Heat Exchangers, etc." No single, comprehensive equipment listing for the facility is available." If the equipment lists are included in the engineering section of the data room, you would direct them to that section in your answer.

Another technique utilized by some consultants hired by a buyer is to attach their checklist to a data request. Where this is the case, it is usually acceptable to generate an all inclusive answer such as "Relating to questions 1,2,3,4,7,8,10,12,16, the information is included in Section XX of the data room." If similar questions are asked relative to one subject, group them together so they can all be answered in one short paragraph.

If a question or data request relates to something not included in the VDD or the data room and the M&A business lead feels it should be, you can develop the answer using the same format as in the VDD report and then add the supporting documents to the data room.

Whether responses to questions will be provided to all potential bidders or only to the bidder asking the question is a decision that will be taken by the Seller's M&A business or negotiating team lead. You should check with the M&A business lead on which approach will be taken prior to developing answers to PS related questions. This could impact the way you frame your answer to a question.

It is normal for all answers to the Q&A to be disseminated to the buyer through the Seller's negotiating team. As such, your answers should go to the person that farmed the questions out to you, and never directly to a buyer's representative. "Shotgun" responses where a string of individuals are copied are to be avoided, unless the negotiating team lead specifically directs that copies should be sent to various individuals.

4.2.9 Reverse Due Diligence

It is appropriate when selling high-risk assets to determine whether the acquiring organization has previous experience and expertise operating such facilities or other means to assure the assets will be operated safely. This is important where the potential buyer is an investment organization and/or the facility routinely utilizes a corporate staff organization that is not included in the staff being acquired with the asset. A Reverse Due Diligence can be done by conducting an internet search on the bidding companies as described in Chapter 3 – Prescreening Potential Candidates. In this case you are trying to ascertain if the potential buyer's organization already has high-risk assets in their portfolio and the level of expertise they have with operating such assets. If possible this should be done once the business team has established a short-list of bidders and before or during the bid review process. If you feel any of the potential buyers may not have the ability to manage safely high-risk assets, you should alert the M&A Project Lead with your concerns and rationale. Conversely if within the short-list you find organizations that have the ability to operate the assets safely, you should pass that information to the M&A Project Lead. This information is important when evaluating the competitive bids and selecting a successful bidder.

4.3 THE ACQUISITION

The philosophy of the acquisition due diligence is to:

- Establish the asset values for the M&A business team taking into consideration any liabilities and risks for each facility;
- Determine the legal compliance of each facility;
- Determine the management systems used at the facilities and potential changes that may be required to align with the Buyer's systems; and
- Gather information for the development of the integration plans.

The level of the Seller's corporate support to the facility also needs to be assessed to assure the facility has adequate buyer corporate support post-closing. Based on the value of the acquisition, levels of significance or materiality (e.g. $100,000, $1Mil, $10Mil) need to be established early in the process to prevent due diligence teams concentrating on nonmaterial issues.

Due diligence as used here is a legal concept that implies the Buyer was provided the opportunity to carryout various investigations to assure themselves of the condition of the assets and business prior to the closing of the deal. As such, if conditions arise after the M&A transaction is finalized that the buyer was made aware prior to the closing, there is usually little if any legal recourse against the Seller unless specifically established in the Sales Agreement. Conversely if an issue arises post closing that was not properly disclosed, the Buyer may have recourse to legal action to recover their costs associated with rectifying that issue. It is important, therefore, that all material issues relating to the process safety aspects of the facility are discovered during the due diligence phase of the acquisition.

Existing liabilities can be handled a number of different ways relative to the acquisition. Where the Buyer is to assume the liabilities associated with any Process Safety issues, the due diligence must focus on defining the various liabilities and estimate the cost of each. These costs should then be fed into the Buyer's business team so they can be included in their economic calculations. An alternate method is for the Seller to retain the pre-closing liabilities. In this situation the due diligence should concentrate on defining the existing liabilities to establish a baseline level of costs. Where the acquisition is of an entire company (e.g. a stock purchase) in almost all cases the Buyer will be assuming these liabilities. The costs of these liabilities, then, need to be included in the business economic calculations since there will be no company after the acquisition to fund any required remediation or indemnify the Buyer.

The future regulatory compliance requirements for the target facility should be investigated at this time. Any pending or proposed regulations that could impact the facility should be captured as part of assessing the facility's current and future compliance profile. Though a facility may be in compliance today, future regulations and proposed changes to industry standards should be considered as they could mean significant future expenditures will be required.

4.3.1 The Internet Search and Initial Data Gathering

Conducting an Internet search on the assets being acquired as discussed in Chapter 3 is likely to develop a significant amount of background information on the target assets or company. The search can reveal red flag issues, such as, negative newspaper articles regarding facility accidents or incidents that should be investigated while reviewing information in the data room or during site visits. These specific issues should be added to your acquisition due diligence checklist to assure they are addressed. It is also beneficial to search the Seller company's external website to determine whether certain documents are available, such as, the company's policy statements, HSE management system and organizational charts (ideally of both the actual operational management of the site(s) as well as the PS related technical support). This type of information can give the PS reviewer an overall understanding of the Seller's organization prior to the actual start of the due diligence activities. Where pertinent documents are found on the company's Internet site, 'official' copies should be requested during the data room Q&A process unless they are already included in the data room. It might be discovered that the documents on a corporate website are not the same documents utilized by a potential target facility. If so the reason for any differences should be investigated either as part of the Q&A or during site visits.

At this point, the possible acquisition of a facility or group of assets by your company may be extremely confidential as a public announcement is yet to be made. In these cases it is likely the company you represent has signed a Confidentiality Agreement with the selling company. Direct contact with the potential target facility staff should not be considered unless approved by the M&A Project Lead. However, Internet searches can be conducted while maintaining the required confidentiality.

During this initial phase of the process, you should ask the M&A Project Lead if they have received an Information

Memorandum from the Seller's organization, and request a copy. While this document may not contain any specific process safety information, it will likely contain a detailed summary of the target assets that will allow you to prepare for the formal due diligence activities.

4.3.2 Vendor Due Diligence Report

The divestment due diligence (also known as a vendor due diligence report as discussed earlier in this Chapter) is often, but not always prepared by the Seller. This Report provides the potential Buyer's business teams summary information that can be used to establish the asset values prior to preparing their initial bid(s) as well as planning out the Buyer's full due diligence. Where the Seller has prepared a comprehensive VDD report, the Buyer's due diligence activities may only need to validate the VDD report. This would include a check for material issues not included in the Seller's VDD report. Where a comprehensive VDD report does not address PS issues it will be necessary for the buyer to prepare an acquisition due diligence report from base documents, site visits and interviews with the Seller's staff.

If a VDD report is prepared by a Seller, it may be distributed early in the process or included in the data room. The PS reviewer should ask the due diligence team leader or the M&A Project Lead for a copy of this document, if one exists, prior to formally visiting the data room. Often this type of document gets held up within the organization of the Seller's due diligence team as part of reviewing it for accuracy prior to being approved for release. In many cases then the VDD report does not get to the Buyer's due diligence team members until the data room is established.

4.3.3 Data Room

The data room can be a physical room or an electronic repository of all the documents collected by the site visit and pre-site visit activities compiled and provided by the Seller to be shared with the potential buyers.

The types of documents usually included in a data room relating to Process Safety are identical to the list for a divestiture included in section 4.1, except for the safety awards and recognition for safe operation from third parties.

Reviewing the data room information is usually done using a checklist of issues or topics that will need to be verified or evaluated. The data room and VDD report should allow the reviewer to begin evaluating material issues related to the assets. Based on a summary of the assets from the Information Memorandum, the VDD and the data room, one of the first questions the PS reviewer should consider is whether all the regulatory and industry standard assessments have been conducted for all of the individual operating units and for the site as a whole. Ideally all such studies are included in the data room. The next questions to answer are what if anything does the information indicate about the present and future liabilities for compliance? As you review the material in the data room utilizing the applicable checklist, areas where no information is provided may become evident. These areas need to be highlighted and noted for further investigation during the Q&A period or while conducting site visits.

The objectives of the initial review of the data room information are:

- To test the completeness of the data provided and establish a degree of confidence in the vendor data provision,
- To identify material issues that may include compliance matters or performance related concerns,
- To establish current or future liabilities for the business team valuation process, and,
- To identify any "red flag" issues that could kill the transaction.

Though you might also want to gather information to begin planning the integration process, you will have that opportunity later in the M&A if your company is the successful bidder.

The potential buyers who are provided access to the data room usually have the opportunity to ask questions or request additional information relating to the business via a formal, written Q&A process. Your questions will likely be submitted to the Buyer's (your company's) negotiating team lead. That individual will, in turn, submit the questions to the Seller's negotiating team lead, who will then direct the questions either to the lead of the Seller's due diligence team or directly to individual Seller due diligence reviewers. The questions and the answers are normally added to the data room for all potential buyers to review. Where gaps in the data room documents are identified, you can ask that such material be provided as part of the formal Q&A process. It is better to review the data room information and ask specific questions or request specific documents, than to ask for all information pertaining to a certain topic and then attach a twenty-page checklist. As noted earlier in the Q&A management section it is possible your questions and the answers to those questions will be shared with all the bidders to a potential acquisition. Whether all bidders are to be provided copies of all questions and all answers or not should be raised with your company's M&A project lead or negotiating team lead. Their answer will impact how you will want to frame any questions.

At the end of the preliminary data room review a report is often prepared discussing the findings of the review as well as identifying key areas for further evaluation. This report is prepared for the Buyer's M&A business or negotiating team as part of their ongoing negotiations with the Seller.

The data room is the repository for all documents shared with the potential buyer(s). As such, additional documents will be added arising from the Q&A and site visits as the Buyer's DD progresses. Depending on the nature of the data room, the new documents will either be added into the index framework of the initial data room or added as a new document file.

At the conclusion of the acquisition, the data room documents should be archived by both the Seller and the Buyer, as they will become critical in the event of any future litigation related to the purchase of the business or individual assets.

4.3.4 Due Diligence Valuation for Bid

As a prospective Buyer your company will submit a monetary bid to the Seller's business team as well as general contractual terms and any conditions or limitations that are felt necessary. As noted earlier in the vendor or seller valuation section of this Chapter, the repair, replacement or upgrade of physical equipment, plant, buildings and control systems in order to bring them into compliance with current standards has from experience exceeded ten percent of the proposed transaction price. While addressing those items can significantly impact the future capital and operating expenditure budgets, the resources required to revise management systems, update or upgrade engineering standards, hazard or risk studies and operating and maintenance procedures should all be assessed as well. This is especially true where the due diligence team finds major gap(s) between the manner by which process safety is managed in the potential acquisition and any regulatory requirements let alone gaps with the Buyer's own approach. To these the PS reviewer will need to add in estimates of the resources and time required to improve or possibly align process safety cultures where necessary.

The Buyer's business team will need to know the liabilities associated with the current operations as well as projected future spending requirements for those liabilities to establish what the Buyer considers to be a fair bid amount, as well as terms and any limitations or conditions. These liabilities and the costs to correct them will be factored into the business team's analyses as part of determining the present value of the business. An integral part of the PS reviewer's task will be estimating the process safety liabilities based on the information provided in the data room and knowledge of existing and pending regulations that could impact the site. A conservative (high) liability cost estimate could adversely impact the bid price. It is usual practice to provide a range of the liabilities and associated costs to address those liabilities. One way to provide such a range is using different probability levels:

- Probability 10% (also known as a P10 estimate) is typically the lowest projected cost where all current and new regulatory requirements are estimated not to have a major impact on the facility or overall business,

- Probability 50% (P50) is typically the cost(s) projected in the Seller's current operating and capital expenditure budgets to maintain compliance with current and projected future regulatory requirements, and

- Probability 80% (P80) is typically the cost(s) where it is considered there are current non-compliance related issues or new regulations that impose stringent new standards on the site or business and the Seller's current spending is underestimated.

It is critical that these valuations are forwarded to the Buyer's business team in a timely manner. The valuations should identify whether or not known corrective PS projects are included in existing budgets to prevent the Buyer's M&A team from making provision for them in their bid when the Seller has already accounted for them. **This information is extremely confidential and should only be shared with the Buyer's M&A business or negotiating team.**

4.3.5 Pre-site Review

Most M&A activity is conducted with a high degree of confidentiality and executed as rapidly as possible. In a sense this is regrettable as history has shown that in the broad process industry sector, rectification of PS related issues identified after a deal closed and integration began contributed to between ten and as much as thirty percent of the total cost of the transaction. Speed and secrecy is unavoidable in today's public markets. However, it is incumbent on both the acquiring and the acquired company to make the PS due diligence process as thorough as possible while taking into consideration business sensitive information.

Ideally, the first step is to conduct pre-site visit interviews of key technical PS staff of the company divesting the assets. A technically competent individual or team from the buyer should conduct these interviews.

The makeup of the PS due diligence teams should be small and consist of competent staff from both companies. In most cases, no more than half a dozen technical people would be required on each side. In addition, each side might consider having an attorney present. Depending on the size of the M&A transaction, the meetings and interviews would not necessarily be full group meetings but breakout sessions focusing on various PS topics or possibly one-on-one's. Special emphasis should be placed on international operations, especially in developing countries. These pre-site visit discussions and interviews can generally be completed in about three days or less depending on the size of the company or assets being acquired. At the end of these meetings or interviews there should be a shared understanding between the buyer's and seller's teams of their collective view(s) even if there is disagreement in certain key areas. To preserve confidentiality, one could conduct this meeting under the umbrella of an opportunity to share and learn from one another's best PS practices. While this approach has been tried many times, anecdotal evidence indicates this strategy preserves confidentiality of the intent to sell only about twenty-five percent of the time.

Once the two PS technical teams complete the pre-site visit interviews they should share their findings with the "in the know" senior executives on both sides who are intimately familiar with the potential M&A. Although each technical team will talk privately with their own management, the value of holding joint meetings with both technical teams and their company's senior management representative(s) in attendance is becoming increasingly recognized. Joint meetings help in assuring both parties are aware of all related issues that will have to be addressed as part of the due diligence process. The senior most operational/management person to whom PS reports in each company should be present in these meetings. That individual can then later brief their respective executives and possibly Board members on key issues that need to be considered as part of the divestment or acquisition.

Having completed the above two steps, the various agreed site visits should be undertaken and completed as soon as possible. Ideally, the Buyer's team will be provided the opportunity to conduct site visits before the bid phase of the process. Very large divestments and acquisitions may involve numerous sites located throughout the world. As a result a representative sample of sites may have to be selected and agreed on, to complete the visits within whatever specified timeframes have been set for the acquisition. This should be agreed and the sites identified in the pre-site interview process discussed above. Site selection is critical in this case. A risk matrix should be developed that not only takes into consideration the actual physical risks of a process safety or major hazard accident but other factors such as the regulatory requirements that in place, actual regulatory oversight, staff demographics, etc. Where this is the case, the potential buyer should clearly articulate to the seller that the buyer reserves the right to add sites not in the sample list. Additional sites visits should be scheduled and undertaken if or where the buyer's team identifies issues it feels could be endemic to the company's operations as a whole.

Prior to the site visits, the process safety reviewer(s) should compile a list of individuals or functional staff they would wish to meet, and submit the list to the due diligence project manager. Also prior to the site visit(s), the process safety reviewer(s) should determine the PPE and other general safety requirements required at the facility. If conducting a DD of a facility where the PPE requirements are less stringent than a similar operation in the reviewer's company, following your own company's PPE requirements is the best option. The buyer's PS reviewer(s) should also request permission to take pictures of the actual site during their site visits as well as whether it would possible to record conversations, meetings or interviews of staff while at the site. Where the seller does not agree pictures can be taken, the buyers should request that any pictures taken of the facilities be provided in the data room or as part of the documentation provided by the site itself. Where the seller does not agree to the recording of meetings or interviews, sufficient time should be built into the visits for the buyer's team to record notes of those meetings.

Prior to the site visit the location of the records should be determined. If various records are not stored at or kept on site, the PS reviewer should request and be provided the opportunity to review those documents wherever they are stored prior to the actual site visit.

4.3.6 The Site Visit and Document Review

The site visit is one of the most important aspects of the due diligence. During the site visit all questions and assumptions made by the PS reviewer(s) need to be addressed, updated and validated.

Since time is of the essence in most process industry M&A's, site visits have to be conducted and completed in a short period of time. The site visit for a large facility usually lasts from three to five days whereas a small facility with few staff may only require a few hours to a couple of days to complete. Depending on the number of facilities to be visited, a group of PS specialists may have to be assembled, briefed and then assigned to multiple M&A due diligence teams.

During the site visits and interviews, questions or concerns not adequately addressed during prior due diligence activities, such as the data room document review, should be investigated. New reports and other documents should be reviewed during the site visit to answer questions the PS reviewer(s) may have regarding the VDD report or the findings of their data room reviews are addressed. If reports discovered at the site are required for the DD process, copies should be requested. As there will likely be a Seller's document management process in place, there may be a formal procedure to request copy of documents that will need to be followed. In an acquisition due diligence, the PS reviewer's duties are not to conduct a formal hazard assessment of a facility, rather the objective is to review and evaluate the quality and types of risk assessments previously conducted and evaluate the overall process safety management systems and culture.

During the actual facility inspections the PS reviewer should conduct the following activities:

- A facility or a plant tour starting with a sit down examination of all the plant layouts and drawings. After this, a cursory plant walk through to get a sense of the housekeeping should occur. Next a detailed examination of P&IDs should be conducted while walking the plant to assess the currency and accuracy of the drawings. The quality of maintenance and upkeep should be checked through oral questioning. Hypothetical "what ifs" should be asked to ensure plant operators are trained considered competent and have participated in drills. High hazard/risk areas should be examined in detail. Questions should be asked about the types of PS related upsets that have occurred in the past five years, how they were handled, subsequently analyzed and what lessons were learned. Any Risk Management Plan submittals should be reviewed in light of what is found as part of the site walkthroughs.

- On the plant design side, facility layout, distances from the fence line of hazardous material storage, location and layout of the main control building, the general design criteria used for blast over-pressure, fire protection and other response systems should all be examined and, if time permits, some of these systems should be tested. The condition of process control panels and their ease of use should be established. Major equipment items such as pressure vessels, tanks, pumps, compressors, reactors and piping should be visually examined to obtain a first order idea of their current state.

- A review of the site's mechanical integrity program and documentation should be undertaken. Random checks of equipment inspection reports made to provide insight on the overall condition of the equipment and effectiveness of the mechanical integrity program. In the area of onsite documentation there should a complete set of all P&ID's for the facility and they should be current and up to date. Further, those who need to be familiar with the P&ID's or parts of them should be randomly interviewed to ensure how well they understand the information in them. The quality and timing of emergency shutdown procedures and training should be ascertained. One weakness of even very well managed plants is their procedure for managing changes (MOC) to equipment and staffing. There is usually a large discrepancy between the theory and practice of how MOC is handled at even the better managed chemical plants. To evaluate the strength of their MOC process, conduct spot checks on how they managed the last few changes and how well they documented their MOC's

- Determine whether reliable and regular training on Process Safety issues is provided and if that training is effective in establishing a robust culture in and among operators, maintenance staff, supervisors and the operational line management. Assess whether staff feel they are empowered to raise and act on process safety related issues. The behaviors and safety culture the senior operators and maintenance staff bring to their job will be a fundamental determinant of overall process safety at a chemical plant.

If the PS reviewer becomes concerned about compliance with regulatory provisions, this should be discussed with facility staff to determine if there are any unique issues or agreements entered into with the regulators in this regard. Often there are special circumstances in a facility's compliance plan the PS reviewer may not immediately recognize. These circumstances should be evaluated by the PS reviewer and discussed in the buyer's DD report. Any non-compliance items discovered by the reviewer should be communicated to the buyer's business team immediately. These could require possible changes to the terms of the sales agreement or correction of the non-compliant issue(s) by the seller prior to the closing of the agreement.

4.3.7 Due Diligence Report and Valuation

The final due diligence report will be developed after the data room document review, the Q&A process and the site visit(s). This final report will likely be an expansion of the preliminary report written after the data room review. Where a preliminary report was developed earlier in the due diligence, it should be agreed with the M&A project lead whether a separate final report is to be produced and the relation of the final report to the preliminary report. In some instances, the M&A team may want only one report - the final report to be written. In other instances, the M&A team may wish to have preliminary or interim reports written so that each successive version of the report demonstrates how issues raised earlier in the process are either closed out or addressed in later reports.

All material issues must be reported to the M&A project lead as the process progresses. The final report will be a summary of the various findings the M&A decision makers need to consider in determining if the issues are of such a magnitude as to 'kill the deal' or whether special provisions will be included as part of preparing their bid.

The buyer will document their concerns regarding what they believe are material process safety issues. The report should also note what constitutes good engineering and management of select items even if they are in compliance. Also the report must identify and quantify variances between current practice in the target sites or business operations and the current practice or policy of the buyer. Very clear and cogent explanations will have to be presented for the conclusions reached as the fate of the M&A transaction may ride on it.

Furthermore, since the site visits would sometimes make-up only a sample of all sites involved in the acquisition, the buyer may need to extrapolate or adjust their findings and estimates to what may be required for the entire company. This should be supported with reasoned assumptions as to why similar issues might exist in those sites not reviewed. The results, for M&A purposes must be translated into immediate and future potential financial or reputation impact to the acquiring company. These estimates could markedly affect the value of the deal.

Undoubtedly, the company being acquired will have disagreements with the buyer's team and their voices must be heard. They, too, should be asked to deliver a very cogent and thoughtful account of why they differ and explain in detail their points of disagreement. For the deal to close an acceptable compromise somewhere between these two views will need to be found. To the extent possible, the Report should assist the executives with that task.

In very large transactions it may not be possible for the pre-acquisition due diligence to include a review of all sites or all relevant documents. In these or other situations where the buyer feels the due diligence could not address all potential issues to the buyer's satisfaction, the buyer should clearly delineate where and what they could or did not cover and identify areas of special concern. Among other things, this could allow the acquiring company to introduce some time-based provisions to be included in the sales agreement. These provisions would allow the deal to go through perhaps with the specified matters or issues being fully absorbed by the acquiring company after a certain period of time, for example three years. Such action preserves the deal yet provides time for the uncertainty associated with specified provisions to be examined and corrected as required.

Upon completion of the merger or acquisition, the entire due diligence process should be carefully documented (possibly with the assistance of an attorney) and preserved in the company archives. In the highly litigious business climate of today, it is important to document key elements of such transactions often no matter what the size of the transaction. This provides for cases where, if necessary, all interested parties can see each step or phase was conducted and documented in accordance with accepted good practice at the time. All areas of significant agreement, disagreement and how they were resolved, as well as what significant actions were taken and how they impacted the M&A transaction should be recorded and retained.

There is no one single approach to effective process safety. All companies address process safety taking into account their own organization's structure, corporate culture, the nature of the business they are in and the risks associated with the various processes and products handled, etc. Some do it better than others. Sometimes the Buyer will have stronger process safety programs, better staffing and a superior safety record than the Seller. The executive management of the Buyer may eventually recognize the programs in the acquired business or facility are more robust or effective and place the correct people in charge to emulate the acquired company practices and procedures. While many of the issues of "how we do it" and "how they do it" are more appropriately handled during post merger integration, these differences should be objectively discussed in the buyer's DD final report.

As a general rule, the final DD report should contain the following information:

- Summarize the assessment process
    - Outline the activities completed,
    - Discuss the activities not undertaken or completed and why they could not be completed,
    - Discuss the potential implications of the items not completed.

- Highlight all material findings
    - Identify and summarize data and information
    - Define process gaps (i.e. differences between 'the way we do it' and 'the way they do it')
    - Identify and quantify any legal compliance gaps

- Discuss recommendations
    - Is this a good deal or not from a PS perspective?
    - Outline the resources/costs/budgets to address issues identified to this point in the process.
        - Describe why there is a need for further assessment and evaluation of the target company's approach to process safety.
    - Discuss the PS liabilities that the target company, operations or sites may have and update the valuation provided after the data room review.
        - Prioritize these issues using some sort of risk matrix.

The DD report will become a historical document. Further a 'good' DD report will assist by forming a baseline for the integration planning process. Although not common it should be recognized that an exceptional due diligence review and report could prevent the need for the resource and time intensive site visits to have to be repeated after the deal closes. At the very least a solid due diligence report can also become the first step in merging the operations of the newly acquired assets into your company's process safety programs.

4.4 DID THE DEAL CLOSE?

A merger or acquisition can close or not for a myriad of reasons including those related to process safety or more general EHS issues, for example ground or groundwater contamination. In the past, petro-chemical companies have even been purchased specifically for some of their clean or safe (from a process safety perspective) assets. Yet it is not uncommon that right after the acquisition a buyer may turn around and divest itself of assets considered non-core or not meeting the overall strategic focus of the new parent company.

CCPS member companies have experienced M&A's where correcting process safety issues entailed costs of between ten and thirty percent of the "fair market value" of a firm being acquired. Yet even costs of this level may not be large enough to stop a really strategic acquisition with significant benefits to both companies. Good financial professionals can surmount these types of differences and construct creative financial instruments and terms rationalizing the deal allowing it to close.

If the deal closes, the business development teams of both parties can inhale deeply and then:

- The Seller can begin archiving data, and
- The Buyer can begin implementing and managing the transition plan.

If an M&A activity was aborted and it was clearly known this occurred because of process safety issues, this could become an ideal case study for your organization, as there could be much to learn from the experience. In these cases, at a minimum you should extract these learnings and revise your checklist to include why the deal failed whether because of process safety due diligence findings, cultural issues regarding process safety, pricing adjustments based on process safety or simply the unacceptability of risk.

5

DEVELOPING THE INTEGRATION PLAN

A 1997 study done by Mercer Management Consulting reported poor post-deal integration was the primary reason that in only 43 percent of over 300 mergers reviewed, the newly merged companies outperformed their peers.

5.0 COURTNEY'S STORY - CONTINUED

It was Monday morning and Courtney was driving to her office. She thought - what a treat to have a real two-day weekend. The first full weekend she had since the CEO had announced their intent to purchase White Hot Chemicals. Every weekend since then she had been away at least one of the two days as part of the due diligence review. The deal closed last Wednesday and was made public first thing Thursday morning. However, that was not the end of it for Courtney. On Friday Carla, her boss stopped into Courtney's office and told her as a 'reward' for doing such a great job on the due diligence side of the deal, Courtney was being appointed to lead the efforts at integrating the approach the newly combined company would take to process safety with Bland Petroleum's. The weekend then, wasn't an altogether respite from the hectic pace of the past few months. Although she did manage to put much of it out of her mind and enjoy the weekend.

But come Monday morning as she drove to work Courtney reflected even though she hadn't identified any one item or issue or collection of issues that were 'deal killers', she knew only about sixty percent of the information she asked to be included in the data room was actually provided by White Hot. Further, their site visits had been limited to just one day at each site. In that day, Courtney's discussions with key staff and tours of the plant provided her with enough information to be aware significant gaps existed between Bland's approach and White Hot's. For one, she found that White Hot had started with two plants, the ones in Texas, and then grew by acquiring all the other plants at a relatively fast rate. She was left with a distinct impression that White Hot was not so much a single company as a loose confederation of individual plants or operations. To a very large extent each of the facilities was allowed to pretty much do their 'own thing' even when it came it to safety and environmental issues. She toyed with an idea of first trying to achieve some level of integration between the newly acquired facilities and then merging them into her company's approach. Courtney was balancing that against integrating them directly when she pulled into the parking lot of her company, parked and started walking to the offices.

5.1 DEVELOPING THE INTEGRATION PLAN AND PROCESS

As noted in the Mercer study cited above, poor post deal integration was identified as the major factor or primary reason that in as high as forty-three percent of mergers, the merged company or operations did not outperform competitors. The Mercer study did not breakdown or provide reasons why this might be the case. However, it is well established that poor planning is often a contributory cause to poor performance. Once a deal has closed there is a need to allocate sufficient time and resources to establish a well-founded integration strategy and plan.

The diagram below outlines the basic process for developing a comprehensive process safety (PS) integration plan. It starts with gaining an understanding and mutual agreement with senior and executive management of the boundaries that will surround the future operation of the newly acquired and combined businesses (i.e., this will serve as the foundation for the overall integration strategy). From there it continues on through a process of review(s) to identify gaps and the extent of those gaps and culminates in the development of a resourced plan for addressing or closing all such gaps.

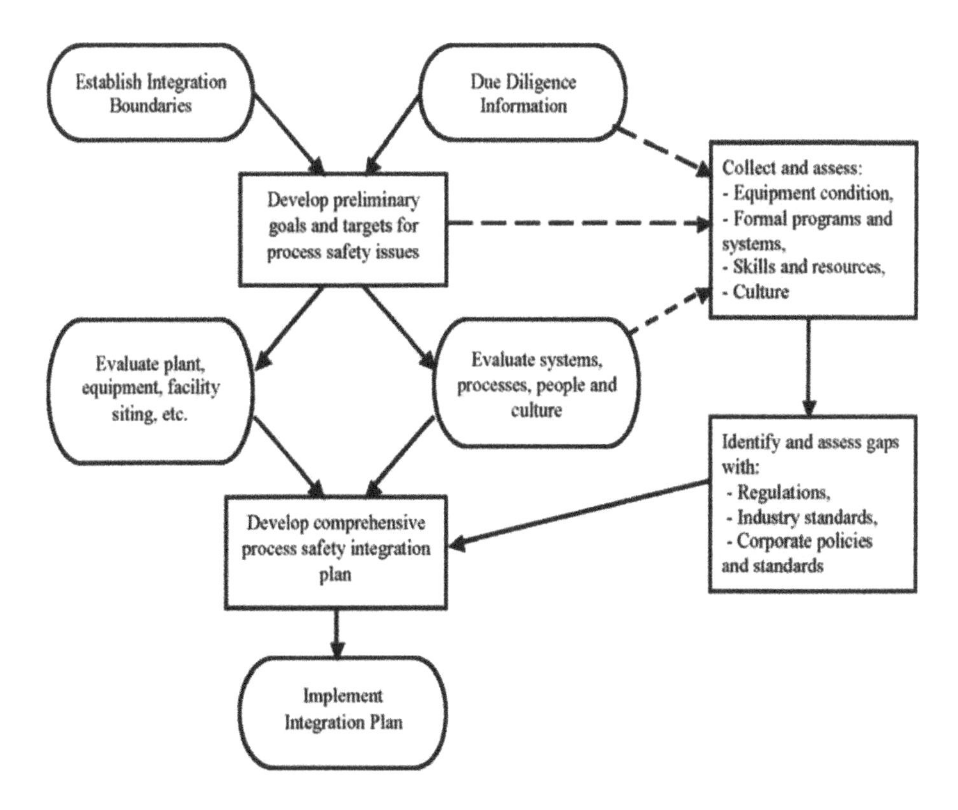

Figure 1. Overview of the Integration Process

In Courtney's case, she had the luxury of having been an integral part of the due diligence team. As a result, after gaining agreement with senior management on the extent to which the newly acquired business or facilities are to be integrated, Courtney might well have the necessary information to develop at least a first order resourced integration plan. However, in many organizations the first time any process safety expertise is brought into the picture is after the deal has closed. In those cases, the process safety integration will have to start anew. When this occurs the first step will be to conduct a series of reviews of the new business and operations and identify whether any gaps exist as well as the resources required to address these gaps.

5.1.1 Step 1- Establishing the Boundaries for the Integration Process (i.e. Establishing the Integration Strategy)

The Process Safety integration lead's first order of business is to gain a clear understanding of the boundaries for the upcoming integration process. Such boundaries might have been discussed at the outset of the commercial evaluation of target assets or companies and again during the due diligence phase. However it is probable after the deal closes, that a new executive or senior management team will be appointed to oversee the integration of the newly acquired business(es). It would be wise to confirm or possibly reconfirm with this team the overall approach your Company will take in managing the newly acquired facilities. The spectrum runs from:

- Allowing the newly acquired business to operate as a self-contained and separate business unit or division,
- Allowing it to operate separately for a period of time and then integrating them into the operations of the parent company both fiscally and at an operating level,
- Providing for a prolonged period (e.g., five to seven years) for full integration. However, it is likely certain processes (e.g., financial, IT, etc.) will be merged much sooner.

- Requiring a quick merger of all systems, operations, etc. (e.g. within a period of two or three years).

Whether to allow the business to run separately or quickly merging them is an obvious input or boundary that will influence the form, nature and resources required to complete the integration process. A business plan for the facility or acquired operations is likely to be developed by the newly appointed senior management team. This business plan will provide further detail and input to the integration plan for the process safety program. The PS integration lead should request a copy of any such business plan or at least those elements or sections concerning the operation of the facilities that impact: the staffing of the unit, IT systems, approach to maintenance, etc.

Historically, many hold the view that to integrate fully, a newly acquired business takes a period of between five and seven years. This is based on the theory it takes five to seven years to change corporate cultures, values and norms. Against that though, there is a growing body of evidence, which suggests the quicker two organizations can be merged, the more effective the integration. Here periods of even three years are seen as being 'too long'. Further, some management teams may originally state and agree that integrating two businesses will take a period of five years. However after about two years these senior managers start to develop a syndrome called 'integration fatigue'. The push will then be on for the integration to be completed within the next one or at most two budget cycles. In other words, when approaching senior management on this issue, be prepared for a decision that sets out shorter timeframes than have been traditionally agreed.

A second bounding condition is how high senior management will want to set the process safety bar for the newly acquired business or facilities. The following three options are commonly considered:

- In compliance with all local, regional and national regulations,
- As above, plus in accordance with all applicable engineering and industry standards in force within the acquiring company,
- As above, plus in accord with recognized best or world-class practice for process safety or major hazards programs and systems.

A reader is likely to ask – What about integrating the newly acquired business' systems to align with the parent company's policies and procedures? A perfectly good question and as it is so important, there is value in making sure the process safety integration lead or team and the senior management integration team have a clear understanding and agreement on this point. Equally important when determining and discussing these points is to get a clear understanding on whether the 'parent' company's systems, procedures and/or culture towards process safety are to predominate or not. Or is there a desire to learn from the newly acquired business and build new systems and a new culture?

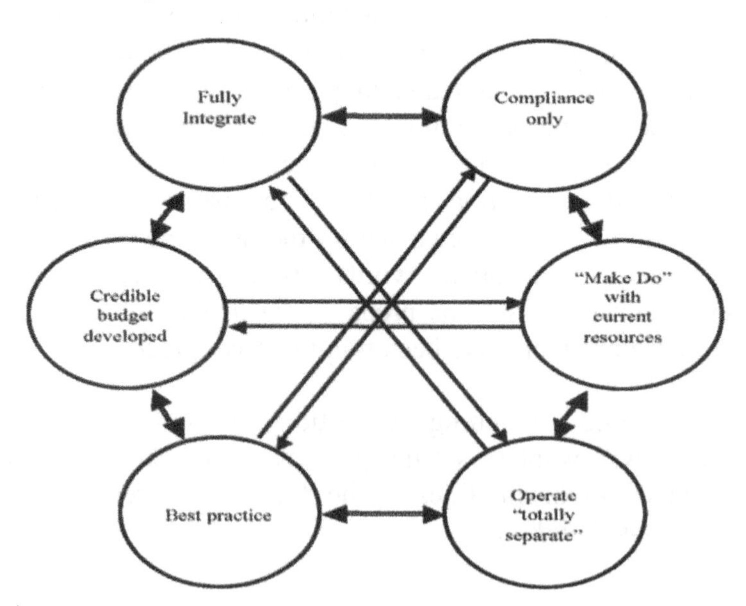

Figure 2. Integration Boundaries

The third bounding condition to be established is whether an actual, and credible, budget will be provided for integration. Or will it be a case of, make-do within current resources and allocated capital and operating expenditure (i.e. CAPEX and OPEX) budgets? Ideally the due diligence process will have identified information on the condition of the plant and physical equipment, the management and IT systems, the staffing of the Process Safety team, timelines, etc. Enough information to build a first order estimate of the potential resources required to achieve the desired level of integration. However, it may be necessary to revisit and reconfirm the potential resources needed to achieve the desired level of integration.

Where this is the case, a possible approach is to use the experience and budgets of the parent company when it was first establishing its process safety management program. These could form a basis for a developing a first order estimate of the budget required. Later though, be prepared to prioritize what needs to be done once the magnitude of the resource commitment is understood. When such compromises become necessary it will prove highly useful to have agreed interim targets or goals for the integration process with senior management. The first such target must be to assure the new facilities comply with any regulatory requirements and from there decide the goals for the next steps.

In summary, step 1 is to develop the boundaries for the overall strategy that will be taken for integrating or merging the process safety aspects of the newly acquired business into the new combined business or organization:

- *The 'extent' to which the newly acquired business or facilities are to be integrated,*
- *The 'height of the bar' senior management is setting for the operation of the new business or facility with respect to process safety, and*

- *The resources and budget that will be committed to achieve integration.*

5.1.2 Step 2 – Establishing the Expectations for the Process Safety Program

There are four aspects to any process safety integration plan:

1. Possible changes or improvements in the physical plant, equipment and the operating process(es)
2. Possible changes or improvements in the human resources that drive and/or support the process safety program(s),
3. Changes to or improvements in the process safety and asset integrity management systems, and
4. Merging and possible revision of the process safety cultures of the two organizations.

Goals and targets need to be set for each of the above aspects. This set of expectations should ensure that:

- All major equipment items (e.g., compressors, pressure vessels, reactors, tanks, etc.) are 'fit for service' and/or maintained in a condition that is considered in accord with RAGAGEP (for example, compliance with relevant API, ASME or NFPA standards, codes or recommended practices).
- Process safety technical support and expertise is adequate to ensure activities critical to maintaining a robust process safety program are in place, sufficient in numbers, and have the authority and resources to administer an effective program,
- Required process safety information such as engineering drawings, material and energy balances, critical operating limits are current and widely available,
- Operating and mechanical integrity procedures and associated training are in accord with current practice and that day-to-day operations comply with the formal requirements of the procedures, and

- Senior and line management, supervisory and technical staff and the operators and craft personnel recognize the importance of process safety, understand their roles and responsibilities in the program and effectively discharge those responsibilities.

Table 1. Establishing a series of measurable Process Safety Performance Metrics for the current state and future targets.

Some Exemplar Goals or Targets	Current Status for Goal	Target(s) for Goal
CONDITION OF PHYSICAL PLANT & EQUIPMENT		
1. Percent of inspections of Safety Critical Equipment completed as scheduled,		
2. Process Safety incidents occurring with a safety interlock in a bypassed condition		
PROCESS SAFETY MANAGEMENT SYSTEMS		
3. Percent of past due or open process safety actions arising from audits, reviews, etc.		
4. Ratio of identified major technical changes in past 'X' years to Management of Change (MoC) reviews undertaken on technical changes in past 'X' years		
5. Ratio of identified major organizational changes in past 'X' years to organizational MoC reviews undertaken in past 'X' years		
6. Percent of Process Hazard Analyses (PHAs) that are 'current'		
7. Percent of Operating Procedures more than one year old or without revalidation		
PROCESS SAFETY STAFFING		
8. Ratio of individuals successfully completing PS training sessions to current number of individuals attending PS training sessions		
9. Percent of individuals completing scheduled PS training sessions in a planned time period		
CULTURAL ISSUES		
10. Percent of staff reporting they feel they have management's support to take corrective actions when a PS condition occurs		
11. Percent of staff reporting they feel the organization takes a pro-active approach to process safety issues		

Many organizations have established a set of process safety key performance metrics for their current facilities and operations. Taking into consideration the bounding conditions for the integration strategy, these established metrics would be a normal starting point for assessing the status of the newly acquired business' process safety programs, procedures, etc. The goals as outlined in Table 1 may prove a useful starting point for setting such expectations for organizations new to process safety who are acquiring or merging with a business or set of facilities with process safety concerns.

A few, perhaps five to eight, such goals or targets should be established for each of the four main aspects of the integration plan (i.e., the state of the physical equipment, the management systems, the process safety staffing and cultural aspects). To the extent possible, the goals should be quantifiable or measurable. This will better assist in determining the breadth of the gaps in the current approach to process safety in the newly acquired businesses against the expectations of senior management. It will also provide the ability to track progress. Where large gaps between programs are found, interim targets as well as a final target for a performance metric might need to be set. Interim targets will help track progress, and more importantly they provide the ability to identify variances or deviations and/or make changes to the integration program(s). [41]

In summary, step 2 is to develop a set of measurable expectations in each of the four main aspects within the boundaries established for the integration strategy of:

- *The physical plant, equipment and operating processes,*
- *The human resources required to drive and support an effective process safety program,*
- *The desired state of the various management systems, programs and procedures that constitute an approach to process safety,*
- *The culture required to ensure these requirements are effectively carried through to day-to-day operations.*

5.1.3 Step 3 – The Process Safety Integration Team

Steps 1 and 2 are focused on establishing strategies, goals and targets of 'what' is to be achieved by the integration process. Step 3 is to identify the resources required to accomplish the required or desired changes.

In some cases, the team that undertook the due diligence process will be assigned to continue and lead the integration process as well. There are certain advantages to this approach one being the due diligence team will have accumulated a body of knowledge of the newly acquired business or facilities that is invaluable to integration. However, not all organizations adopt this approach. In fact generally, once the due diligence is completed, an entirely new team will be assigned responsibility for integration. In many cases, commercial evaluation and due diligence are undertaken by a specialist group within the company's business development division. The focus of this group is often to identify and complete the due diligence through to closure of the sale. Once the deal is completed, this team will be charged with identifying other possible acquisition targets or candidates. Where this is the case, integration of the newly acquired business is transferred to the senior management team of the operating business unit into which the newly acquired business will be merged.

A second condition that will shape the size and nature of an integration team will be the budget senior management allocates to the integration process. If the integration is to be accomplished without being provided specific time, resources or budget, the integration team could take the form of a 'virtual team'. In other words a team formed from the current process safety resources of both the newly acquired businesses as well as the parent company. As a result, establishing a dedicated team with possible assistance from external resources, where required, is unlikely to be approved by the senior management integration team.

The integration process has the potential to be highly resource intensive. The ideal is to obtain approval for establishing an integration team dedicated to achieving a successful integration or merger. When forming such a team, the opportunity to include resources and expertise from the newly acquired business should not be overlooked. One benefit of such an approach is obtaining buy-in from these key individuals to the possible changes that will occur as the two approaches to process safety are merged together.

The composition of the integration team is likely to change as the integration process moves through two distinct phases of first identifying the gaps between the two approaches and then secondly closing those gaps. While the technical skills required between the two phases are likely to be similar, the experience in auditing and reviewing during the first phase versus developing and building systems in the second will change.

The first phase will be focused on establishing the current state of various programs, processes and the physical plant and equipment. The integration team members then, should be well experienced in reviewing management systems, in evaluating the condition and state of the physical equipment, the capabilities and experience of staff and last but not least, safety cultural issues. In this first phase, the technical skills of mechanical and/or process engineering, process safety engineering, management and management systems, etc., will need to be balanced with experience in evaluating, reviewing and auditing the associated technical issues.

In the second phase, the same sets of technical skills are likely to be required depending on what gaps are identified in the first phase. However, in the second phase those technical skills must be supplemented with experience and skill in improving, creating, and implementing changes or perhaps instituting completely new programs, systems, procedures and cultural shifts required to meet management's expectations.

Table 2. Possible skills the Integration Team(s) will require

Possible Skills Required	**First Phase - Evaluation** (auditing & reviewing experience)	**Second Phase - Development** (development & implementation experience)
• Asset or Mechanical Integrity	X	X
• Maintenance and Inspection Planning & Scheduling.	X	X
• Hazard & Risk assessment	X	X
• Human Factors		X
• Management system program development		X
• Procedure writing		X
• IT software systems		X

How large of a team or team(s) will be required? There is no one answer to this question. In acquisitions and mergers with multiple sites, there will be a need to identify and mobilize multiple teams. Multiple teams may be required both to complete the process within a reasonable period of time or, at least within the deadlines set by senior management, as well as cater for differences in regulatory regimes in different countries or regions, language differences, etc. Ideally you should plan for one integration team to be assigned to each of the individual sites acquired.

How large should each team be? Each team will need to have the skill sets as outlined in the Table 2. Generally this might mean at least one individual with the proper skills and experience in each of the main aspects (i.e., one for the physical equipment, one for the management systems, one for the staffing and one for the cultural issues). This should not be interpreted to mean the skilled individual assigned to each of these particular areas would actually do all the related work in that element. Throughout much of the integration, that individual will be identifying, mobilizing and overseeing the work of others to accomplish all the potential tasks associated with integration.

Assembling a team or possible teams provides an excellent opportunity to set the future tone for the integration process. For an integration or merger process to be successful, there is a need to establish buy-in to the newly combined way of operating the business(es) at all levels. Including key staff and individuals from the newly acquired business on the review teams presents the opportunity to start on-boarding them into the future approach to process safety. While process safety expertise is a vital part of the team, the team should also include line managers and other technical staff who are vital to the success of a process safety program (e.g., engineering or project management). Where it is determined specialist skills such as experience with implementing new IT systems are needed, these should be appointed to the team on an ad-hoc basis.

Finally, when planning the size, number and composition of an integration team or teams take into consideration that the energy demands on these teams will be considerable. The potential for team members to 'burnout' is high. Further team members may be lost by being appointed to positions whose demands prevent them from continuing to be a part of the team or attrition. The process safety integration lead needs to plan for such eventualities and have alternates or a mechanism in place to rotate members of the integration team(s) as and when the need arises.

In summary, step 3 is to identify and form an integration team:

- *The team(s) will need to have technical skill and experience in each of the four main aspects of an integration (i.e., the state of the physical plant or equipment, the management systems, the human resources and cultural issues),*
- *During the first phase (i.e. that of establishing the gaps that exist in each of the four aspects) the team will need to have*

skill and experience in auditing and reviewing such systems and programs,

- *During the second phase the team members will need to have skill or experience in developing and implementing systems, programs and procedures,*
- *Be prepared and plan for team members to be rotated or replaced whether as a result of burnout or potential advancements in their careers by taking other positions.*

5.1.4 Step 4 – Assessing the Gap between the Current Approach and Expectations

The next activity is to establish a baseline of actual process safety practice within the newly acquired business' various sites or facilities. In an ideal world, the due diligence reviews should provide sufficient information to establish a baseline of:

- The general condition or mechanical integrity of the physical plant and equipment with a focus on the degree of compliance with all regulatory requirements or to a good industry standard,
- Whether the skills and technical expertise are currently in place to manage a process or major hazard safety program and the degree of internal versus external support used to support and maintain such programs,
- The state of the various management systems in place and form the framework for the newly acquired business or facility(ies) process or major hazard safety program, and
- The culture of the management, supervisory levels and operating and maintenance staff towards actual day-to-day adherence to or implementation of an effective process safety program.

However, it is often the case that the due diligence team did not have the time or possibly access to the necessary information to establish a robust baseline. In these cases, the PS integration team will have to carry out a series of reviews of the operations of the newly acquired business or facilities to develop an implementation or integration baseline for each of the above four major aspects.

Where it proves necessary to undertake such a review or reviews, the usual steps that will need to be completed are summarized below:

1. Develop a review protocol,
2. Assemble a skilled and competent team or teams,
3. Orient the teams in the requirements of the review protocol,
4. Prioritize the order that sites or facilities will be reviewed,
5. Make provision to provide senior management with periodic updates especially of any critical issues requiring immediate attention,
6. Carry out the reviews,
7. Compare the results to expectations and develop a 'Gap(s) Report'.

A review protocol will first need to be generated. A starting point for the development of this protocol are the expectations and process safety metrics as discussed in Step 2. The need for such a protocol is especially important where the size of the new acquisition means multiple teams are likely to be mobilized. Such protocols help maintain a level of consistency of both the activities as well as the findings developed when multiple teams are mobilized. When generating this protocol the boundary conditions established in Step 1 must be kept in mind at all times. As an example, if the decision is to operate the facilities as a separate stand-alone business, compliance with appropriate regulatory requirements will be mandatory. However, revision of various procedures, guidelines, engineering standards in order to bring them into compliance with those of the 'parent' company, would not form a part of the review.

In addition to the metrics in Table 1, the checklist of process safety issues contained in Appendix A of this Guideline is a useful reference for developing a baseline review protocol.

In parallel to developing a review protocol, the priority for reviewing the various facilities or sites should be determined. The order or priority for assessing sites should be based on any perceived risks that a particular site or operation might have a major hazard accident or that the risk of a serious breach of a regulatory requirement exists. Here, the members of the due diligence team can assist with developing a picture of any potential risks at the individual sites. They will have formed an opinion of the general state of the physical plant, the overall nature of the technical skills and expertise currently available and the state of various management systems from their review of various information provided in the data room and visits to the various sites. As a result, their input into prioritizing which sites should be visited first will prove invaluable. However, the practicalities of day-to-day operations, geographical location, accessibility to the plant due to environmental factors (winter, storms, etc.), will have to be merged with the perceived level of risk when establishing a final review schedule.

Now comes a very bold step to consider. Are you willing to allow a combined team of newly acquired business and parent company staff to review a representative set of the 'parent' company's facilities as well as the sites and operations of the newly acquired business? And, as important, will you commit that whatever the review team finds with respect to gaps in the parent company facilities, that those gaps will be treated and addressed just the same as any gaps found in the operations of the newly acquired business? The first response to such a suggestion is likely to be a grimace and retort to the effect of '...that will unnecessarily add to the time it takes to complete these reviews.' The answer is yes, of course, it will add a few weeks. However, the investment in time and resources of a few weeks will achieve a much greater return in the form of a more committed albeit small group of staff from the newly acquired business to making changes to their approach to process safety where necessary.

Provisions should be made to apprise senior management periodically of the status of the baseline review. Of importance is to inform senior management of any critical risks identified, along with possible temporary 'patches' that should be implemented to correct the deficiencies. The overall form and frequency of an interim report should be agreed with the senior management integration team as part of the initial planning phase. In general, these reports need to be timely, brief and provide estimates of the associated costs required to address any issues raised in the report. Such reports are sometimes referred to as 'emerging issues' reports. In addition to identifying these emerging issues, it should also be agreed with the senior manager or senior management integration team of the possible need to develop a ninety to hundred day corrective action list. This list would be restricted to critical and high-risk items identified as part of the baseline review. In other words risks or situations the parent company would classify intolerable, therefore requiring some form of immediate corrective action.

The goal of this review is to establish if any potential gaps exist between the current approach to process safety in the newly acquired assets and the desired approach as well as to assess the extent of those gaps. As part of the review process then, the team must be capable of not only identifying potential gaps but also to provide a measure of the distance between the current state and the intended state to be taken towards process safety.

Figure 3 provides a qualitative means for depicting the gaps between a rule based and risk-based approach to safety. It further contrasts the overall management styles organizations often take. If in a particular merger or acquisition one organization has a very centralized approach to management and is very rules based, versus the other organization having a decentralized approach and a very risk based approach, considerable time, resources and management effort will be required to bring such approaches to a common ground.

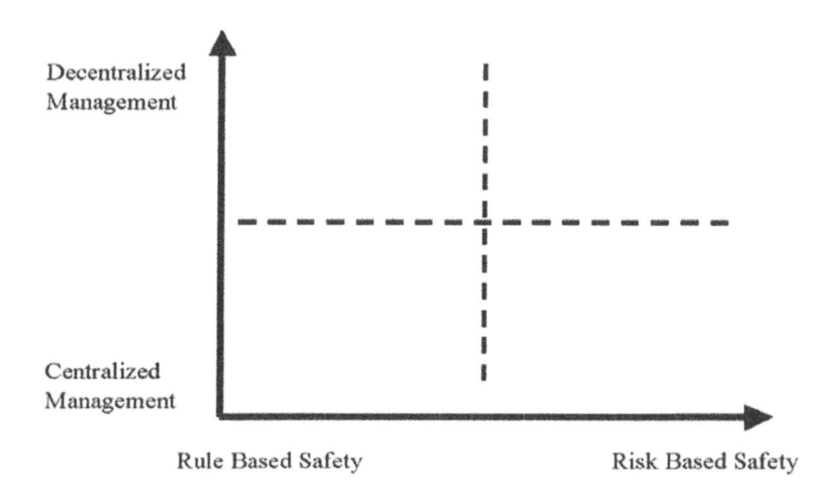

Figure 3. Assessing the gaps between the current and desired states

The size of the acquisition, number of sites or operations, their geographical location(s) will all impact the period of time required to complete such a review and establish a baseline. However even for a 'mega-merger', senior management will likely want the review completed in a period of months (e.g., three, six and at the outside nine months) rather than over a period of a year or more. Ideally, this review should be timed and completed to incorporate the results of its findings into the next CAPEX and OPEX budgetary cycle of the parent or newly formed organization.

Step 4, in summary:

- *Establish the expectations for the newly acquired business.*
- *Use data and information from the due diligence phase to establish this baseline, or if necessary develop a protocol of issues to be examined, form an interdisciplinary and inter-company team to undertake a series of reviews of each of the sites or facilities,*
- *Identify and quantify potential gaps between the current and future desired state of:*

  - *The general condition of the physical plant, equipment and process operations,*
  - *The process safety related management systems or programs in place,*
  - *The technical skills and resources that are employed to drive or support the process safety programs,*
  - *The culture towards process safety of the management, technical and supervisory staff and the operators and maintenance staff,*
- *Develop a gap analysis or report between the current and desired states,*
- *Complete this review in a period of, at the very outside, nine months.*

5.1.5 Step 5 – Developing the Action Plan

The action or integration plan will need to address:

- Hardware - for example; buildings, physical plant and equipment, and instrumentation,
- Software - for example; mechanical integrity programs, procedures, process hazard analyses (PHAs), risk assessments, the state of records and information, and training,
- Human resources related issues including; technical skills and experience required, reporting relationships, and authority and responsibilities,
- Culture towards process safety.

Some organizations may wish to combine the cultural issues into the human resource aspect of their integration plans. Whether culture is dealt with as a separate issue (recommended) or as part of the human resource aspect of the plan, its overall importance to assuring the formal or written process safety requirements are effectively imbedded into the day-to-day operations needs to be recognized, properly addressed and resourced.

Make Provision for Addressing Intolerable Risks

As noted earlier in this chapter, the baseline site visits may identify what the parent company considers to be intolerable risks. Such situations may be discovered as a result of:

- Observations of the state of the actual physical condition(s) of tanks, reactors, heat exchangers, piping or pressure relief valves by the review team,
- No records, or records showing various tests, inspections, or maintenance of such equipment are well overdue,
- Reviews of records or discussions with operators and maintenance staff of near misses, incidents and leaks.

Where such situations arise, there will be a need to obtain further information on the actual state of the physical plant, equipment, operations, etc. This is necessary to assess better, what the existing condition(s) are and from that develop a plan of corrective actions. This may entail mobilizing specialist inspectors, instrumentation technicians and possibly mechanical or chemical engineers to investigate the situation(s) and make recommendations that can be implemented in the near term to reduce any unacceptable risks.

The Physical Plant, Equipment and Process Operation Plan

This plan is that part of the overall integration process most likely to impact current and future capital works budgets (i.e., the CAPEX budgets or funding). It may highlight the need for improvements in the overall site lay-out, spacing between process trains and/or critical equipment, the physical condition of the processing plant or equipment and critical utilities, the instrumentation and control systems, relief systems or fire detection and protection.

Table 3. The structure of a possible Capital Improvements Plan

Possible Structure of a Capital Improvements Plan								
Plant or Issue to be addressed	**Respon-sible Party**	**Budget Esti-mate**	**Status & ETC**	**Schedule Months**				
				M1	**M2**	**M3**	**...**	**M'X'**
Address location of occupied buildings								
- Conduct Facility Siting Analysis (FSA)								
- Obtain estimates to address FSA findings								
- Obtain Authorization to implement changes								
- Implement Corrective Actions								
Upgrades to control room, move work-shops, etc.								
⋮								

In many cases, as identified in Table 3, the first activity will be to complete a detailed study or analysis of various situations or gaps identified in the baseline review. Where this is the case, it is possible the integration planning team will not have sufficient information to provide the senior management integration team with a detailed estimate of the costs needed to correct the various gaps. At the very least, the process safety integration lead or team will need to provide senior management an estimate of the time-frame and costs to undertake any further studies required. One possible starting place for developing such a first order budget is to use the 'parent' company's own experience for completing such studies. Remember, though, the 'integration clock' is ticking against a two or three year outer limit senior management will have in its mind. And that upwards of six to nine months may have elapsed in completing the reviews of the various facilities in a large or 'mega' acquisition. Also, there will be the pressures of where you are in your company's budgetary cycle. All these will be driving senior management's desire to receive some sort of monetary estimate from you on any necessary capital works that will have to be undertaken. As a result, you are likely to be under considerable pressure to complete any additionally required studies and from these, work-up a budget estimate as quickly as possible.

Individual and departmental responsibilities' for various parts of the capital works integration plan must be agreed with the senior management integration team. The process safety integration leader or team will most likely be responsible for completing certain studies (e.g., facility siting, process hazard studies, relief system studies). In those cases, the PS integration lead will need to be provided sufficient resources and budget to complete that phase of work. However, after completing this investigative phase, implementation of the proposed corrective actions should become the responsibility of the site or perhaps a technical specialist team or group (e.g. an operations excellence team) within the new organization. The process safety integration leader is likely to retain a responsibility for monitoring progress in completing the various actions and providing the senior management integration team with periodic reports on progress but not necessarily retain responsibility for implementing these corrective actions.

The Human Resources Plan

This plan will need to address three aspects of the resources employed to drive and support the overall approach to process safety. These include:

- An orientation and on-boarding process for identified key individuals who will be major stakeholders in the delivery of the future process safety programs,
- Rationalization of the future process safety functions, including organizational responsibilities and reporting lines,
- Aligning and imbedding the necessary culture in both the newly acquired business as well as the combined business.

Differences between process safety policies and major program elements will have been identified as part of the gap assessments carried out in the baseline reviews. The nature of orienting key individuals into the future approach to be taken will depend on whether the strategy is that the parent company's policies will predominate or there is a desire to learn and merge the best of the respective approaches of the previous separate organizations.

Where the strategy is one of learning and merging the best elements of each, there will be a need to pull together a team of individuals from both organizations, extract the best elements and develop those into a new policy for senior management's approval. Such a team should have a senior manager(s) actively participating to facilitate eventual approval by the senior management team as a whole. Once the new policy is developed, a communication and rollout of the new strategy will need to be developed and implemented. All of this should occur within a period, at the outside, of six months following closure of the acquisition. On the other hand, if the strategy is that the parent company's policies will predominate, there is already likely to be a system in place for orienting various levels of the management, supervisory staff and key technical staff into the requirements of those policies as well as the structure of the overall process safety program. In these cases, this phase of orienting key individuals in the newly acquired business should be completed within a period of ninety days from close of the deal.

In addition to differences in the policies and major program elements, these initial orientation or on-boarding programs need to address:

- Differences in corporate communication styles, and
- Perceived or real differences in the approach taken to HSE issues, in general, as well as the process safety aspects of those programs.

In most acquisitions, senior management will be looking for 'synergies' in the two organizations. In plain terms, areas where they feel they can combine and reduce the number of staff to manage support and operate the combined businesses. The support and delivery of the process safety program(s) is as likely to be subject to identifying such synergies as any other part of the newly combined business. It is probable then within the first ninety to one-hundred twenty days after the deal is closed, the senior management integration team will be asking the process safety integration lead or team for a plan to rationalize the process safety related operations. This plan will likely need to address:

- Organizational responsibilities and accountabilities for process safety matters,
- Rationalization of the organizational chart for process safety,
- Clarification of functional positions and responsibilities for process safety,
- A strategic staffing plan for process safety related personnel that addresses rationalization of process safety related staff

When developing such a plan it is important to note that within European Union countries, sites that must prepare a 'Safety Case' have requirements to undertake an assessment of changes to the organization that might impact the manner by which major hazards are controlled and managed. This is also becoming a standard practice in and among companies in the U.S. that fall under the PSM/RMP regulations. As an integral part then of these changes, there is a need to develop a Management of Organizational Change plan. This plan will need to examine whether, after making the intended changes, the risks of a process safety or major hazard event are still controlled to acceptable levels. As well, this Management of Organizational Change assessment will need to identify and address any risks associated with planned changes to the management systems and procedures.

The timeframe for identifying and implementing these synergies is likely to be set by the executive management of the newly combined business or the senior management integration team. There is every possibility they will require the synergies to be realized prior to the next budgetary cycle of the company. This could mean, work on identifying and implementing synergies will have to be completed in a period of as short as six months assuming a mid-year closing of the acquisition or possibly as long as eighteen months.

The capital works requirements associated with making required changes to the plant and equipment are likely to receive significant management attention because of the transparency associated with their being line items in capital budgets. In addition changes to the management systems are likely to place significant demands on the human resources allocated to support and drive the process safety programs. Unfortunately management's attention on the CAPEX impacts and human resources implications often means cultural issues are likely to take a backseat to these other issues as the integration process progresses. This despite study after study that point to differences in cultures sitting at the heart of why many mergers and acquisitions fail to return the value expected of them. Further, cultural issues are so entwined with other policies, programs and processes it is practically impossible to extract out a sub-element such as process safety and isolate it from other changes that are occurring. Factors such as: changes in signature authority; reporting relationships; and whether the acquired business approached safety from a risk or rules based process, will all impact on the overall culture currently instilled in the staff you are trying to change.

Establishing a positive culture to process safety changes can become especially demanding where the decision is made that the newly acquired company is to be fully integrated into the parent company's policies, programs, management information systems, technical standards, etc. In these cases, many in the newly acquired business can develop a view that these changes are a reflection on their own past work and activities being '...not good enough'. Where this becomes widespread, the risk is high of significant gaps developing between the requirements of the new formal process safety programs and how day-to-day activities are actually performed. Engaging staff in meaningful discussions and communications at all levels and across all disciplines is vital in preventing a wholesale rip from occurring between the cultures of the two businesses being merged. Meaningful must include actively acting on the suggestions of staff and working with them to help them understand and accept why in certain cases their suggestions cannot be implemented. This means line managers and technical support staff will have to set aside the time to engage in such discussions to listen actively and respond empathetically where required. They will also have to actively follow up on any promises made arising from such discussions.

Cultural change takes time. Further, it is a process that must be actively linked to other changes being implemented as the integration process moves forward. Simple changes can have effects on the culture that are significant. Changes in how operating or maintenance procedures are formatted, the implementation of new reporting procedures, forms and requirements, changes to IT systems from one Computerized Maintenance Management System to another, all will impact the culture of those directly involved with such changes as well as those they routinely interact with. Making such changes without assessing the potential consequences on the culture and behaviors of the staff will likely result in these new systems not returning the value or benefit intended. Worse it might well spur the development of home-grown workarounds that not only defeat the purpose of the intended change but do not provide the ability to accurately track the current state of a critical process safety related system.

'Software' or Formal Process Safety Programs, Procedures, Systems

This part of the integration plan, covers a wide spectrum of the formal programs, systems and practices that are in place within the newly acquired business, including:

- IT systems, procedures, processes, forms, etc. for reporting incidents and accidents,
- Engineering or technical standards for plant and equipment, including standards for inspection and testing of equipment to assure it is 'fit for service',
- Maintenance, inspection and tests records and systems are in place to capture and track such information,
- Management of technical and organizational change,
- Operating and maintenance procedures and programs are in place and kept current,
- The nature, form and completeness of all required process safety related information, etc.
- The procedures for undertaking PHAs, the quality of the most recent PHAs, and the mechanisms for tracking and closing out findings and recommendations.

From the above list, it is easy to see that this section of the plan can grow to become a very resource intensive part of the integration process. It is unlikely to place much demand on future capital budgets save where it is decided there is a need to implement new IT systems to support these various activities. Depending on the nature of the gaps between the current policies, systems, procedures, processes, practices, etc. and the expectations for these activities, considerable effort in the form of rewriting such documents, undertaking any missing studies, recreating or creating required records could place significant demands on internal resources or require budget to bring in outside skills and expertise to supplement internal resources. As resourcing this part of the integration process will become an issue, the gaps should be prioritized using a risk matrix that considers not only the potential physical risks as well as the risk of a regulatory action, risks to the corporate reputation, etc.

There is a bit of a circular or iterative process between establishing the gaps in the actual 'software' elements of a process safety program and the activities undertaken to establish the status of various hardware or equipment items. As an example, in trying to assess the status of pressure vessels, tanks, piping, etc. the review team will want to start with examining the state of the current records on such items. When gaps are found in those records, it will dictate the need to undertake various required inspections, tests, etc. to determine whether such equipment is currently fit for the service demanded of it. As a result the team will have identified not only a gap in the state of the current records, but in the systems that should be in place to assure such information is collected, collated, analyzed and tracked in accordance with good industry practice or RAGAGEP. However, the gap in this information could be the result of a gap in the formal program or from the application and implementation of what is contained in the documented formal programs. Depending on what is found, it could mean the integration process will have to address and revise the formal programs or possibly the actual day-to-day practices to bring them into accord with the requirements of the formal programs.

Out of this laundry list of possible software related changes, it is not unusual that organizations will want to merge their accident and incident reporting systems very early on in the integration of the two businesses. It is one issue that the senior management integration team is likely to give high priority. This frequently means the IT systems used by the newly acquired business to capture, collect, collate and report these issues as well as the form and format of any reporting requirements will need to be replaced. Just this step alone could take a period of six months to a year to fully imbed the new practices in and among the various sites and facilities acquired.

In developing this part of the integration plan, there is a need to establish what priority the senior management integration team places on addressing various gaps that exist in these programs and systems. It is most probable where there are gaps in the current information such as: records of tests and inspections, none or inadequate PHAs or facility siting studies, inadequate records of near misses, incidents or accidents, etc. the near and interim term focus would be on undertaking the necessary studies to better evaluate these gaps. While a stopgap approach, it is a realistic one. One that recognizes there are always constraints with respect to the resources available to undertake all the changes necessary for a successful integration and merger. Having addressed all critical information gaps, one can then start to implement the changes required in the governing procedures, processes and systems to ensure this information is kept current and in accord with established expectations.

The Consolidated Integration Plan

Whether the senior management integration team requires a consolidated plan or not, it is a valuable tool for the process safety Integration team or lead to have for tracking progress of the individual streams and perhaps as importantly identifying potential synergies, critical paths, resource limitations, and interactions between the various elements or streams. structure of a consolidated integration plan might take the general form shown in Table 4:

Table 4. A possible structure for a consolidated Integration Plan

Possible Structure of a Consolidated Integration Plan									
Issue to be addressed	Respon sible Party	Budget Estimate	Status	Schedule Months					
				M 1	M 2	M3	...	...	M' X'
Establish and Mobilize P.S. Integration Team(s)									
'Make Do' or Budget agreed									
Identify a PS single point of contact in each site									
Identify interfaces with HSE integration team									
First order estimate of required skills for each team									
Etc.									
Carry-out site reviews and estimate gaps									
Establish PS Performance Metrics									
Develop a review protocol									
Train review teams in use of protocol									
Prioritize sites for review									

Possible Structure of a Consolidated Integration Plan									
Issue to be addressed	**Respon sible Party**	**Budget Estimate**	**Status**	**Schedule Months**					
				M 1	**M 2**	**M3**	**...**	**...**	**M' X'**
Carry out site reviews									
Periodically report findings to senior managers.									
Collate findings identify & measure 'gaps'									
Develop plan(s) & budget(s) to address gaps.									
Identify 'intolerable risks' & develop a 90-100 day corrective action plan									
Physical Plant & Equipment (i.e., capital works plan)									
Identify the need for additional studies, inspections, tests, etc.									
Facility siting studies,									
UT tests of piping, tanks, pressure vessels, etc.									
Relief and blow-down systems,									

Possible Structure of a Consolidated Integration Plan									
Issue to be addressed	**Respon sible Party**	**Budget Estimate**	**Status**	**Schedule Months**					
				M 1	**M 2**	**M3**	**…**	**…**	**M' X'**
sizing, testing, etc.									
Etc.									
Human Resource Plan									
Develop RACI chart for PS									
Establish communication channels									
Develop & implement new organizational structure									
Etc.									
Management Systems Plan									
Revise & merge strategies and policies									
Revise & merge governing manuals & programs (e.g., PS manual)									
Address issues with operating, maintenance, inspection and testing procedures									

Possible Structure of a Consolidated Integration Plan									
Issue to be addressed	**Respon sible Party**	**Budget Estimate**	**Status**	**Schedule Months**					
				M 1	**M 2**	**M3**	**...**	**...**	**M' X'**
Revise and implement new incident-accident reporting requirements and procedures									
Etc.									
Cultural Issues									
Develop programs and orient & 'on-board' key individuals (e.g., line managers, key technical staff, etc.) into new approach									
Align communication style & systems to support desired culture									
Roll program out to supervisory staff									
Engage operators, maintenance and key contractors									
Etc.									

Some of the CCPS member companies have developed guidelines born out of their experiences, which might help when trying to estimate the time, resource and budgets required to complete a successful integration:

- Language – Where a single or number of newly acquired sites are located in a different country and their documentation is in the language of that country, you might want to add another thirty percent into your resource estimates to account for this difference. If, however, the documentation is in the same language as the parent company and the bulk of the workers are bi-lingual you may only need another fifteen percent of time to account for this fact.
- Regulatory Climate and Culture – Where there is a major difference between the regulatory climate under which the parent company operates and the newly acquired sites, again you might need to factor in an additional thirty percent of time and resources to address these differences. Where the perceived gap between regulatory climate and safety culture is small, it might require an additional fifteen percent of time to address and reconcile such gaps.
- Support – If the newly acquired business or sites are distant from or in a country where the parent does not have a current operating presence, add in upwards of thirty percent additional time and resources to account for mobilizing the necessary resources to these various sites. If there is a current operating presence in the country or region, you might still need to factor in another of fifteen percent additional time and resources to support the required changes.

Step 5 – In Summary:

- *Develop a program to address and put into place temporary controls for all risks deemed 'intolerable' within a period of ninety to one-hundred twenty days,*

- *Develop a capital works program to address deficiencies or gaps found in the physical plant, equipment and processes.*
  - *Responsibility for completing these works is likely to rest with groups or departments other than the process safety group*
- *Develop a plan to rationalize resources and organizational responsibility for process safety related activities,*
  - *As part of this plan address the need to establish the desired cultural changes required to assure process safety is imbedded into the day-to-day activities of the newly acquired business.*
- *Develop a plan to address gaps in the policies, procedures, and systems that govern the desired approach senior management wish the new organization to achieve.*
  - *Where there are gaps in the state of current information on the physical plant, equipment and processes the plan should address and correct these gaps first,*
  - *Once these gaps are addressed, undertake the necessary changes to address and modify the process safety governing processes and systems.*
- *Develop a consolidated plan to manage the overall integration process, track progress towards goals, and implement corrective measures as soon as variances are identified.*
- *Develop a Management of Organizational Change plan that assesses whether the planned changes will still control risks to acceptable levels and identifies the controls necessary to assure such risks continue to be controlled.*

6

IMPLEMENTING THE INTEGRATION PLAN

"There is nothing more difficult to take in hand, more perilous to conduct or more uncertain in its success, than to take the lead in the introduction of a new order of things."
Niccolo Machiavelli – The Prince

6.0 COURTNEY'S STORY – CONTINUED

It was just passing 6.30 PM. Courtney was sitting, alone, in a conference room of one of the Texas plants her company had recently acquired. She was staring, blankly, at the various flipchart pages hung around the room. The only thought that kept going through her mind was - did we really just spend the past six hours discussing whether to call a number of recently revised documents - guides, guidelines, procedures or standards? Surely it can't be that difficult to settle on an agreed set of terms for such material? There weren't these issues in Bland. And how am I going to explain this to the senior management integration team? They are expecting us to get through this by the end of the year. At the rate of this meeting we'll be lucky to finish revising and rewriting the various standards, procedures, etc. in five years!

Any integration process involves change. To complete successfully the integration of two possibly disparate approaches to process safety requires building momentum in and among various managers, teams and individuals of the newly acquired business to accept change. Without the desire to change, the probability of revising the management systems, modifying equipment or processes, in line with senior managements' expectations is remote. If you have not read Chapter 5 yet, we suggest before continuing with this Chapter to take the time to read Chapter 5. If you have read Chapter 5 and have developed an integration plan, implementation entails the mechanics of working through that plan. However, having a good plan does not automatically guarantee success. A study by KPMG of change programs found 80% of all such programs failed to achieve the stated objectives for those programs. Before discussing some of the mechanics of integrating two approaches to process safety there is value in examining the issue of 'change' and why planned changes often fail to meet expectations.

6.1 A GENERIC CHANGE MODEL

A number of models for change exist. The following discussion is adapted from a change model developed by John Kotter (Professor, Harvard Business School). It was first published in 1996 in his *Leading Change*[42] book. In 2002 Professor Kotter and Dr Dan Cohen who together had further studied change initiatives published their findings in the *The Heart of Change.*[43] Kotter and Cohen's work focused on trying to identify the root cause(s) of why as reported by KPMG, 80% of all change programs fail to meet initially stated objectives. Arising out of their studies Kotter and Cohen found:

> *"More than any other single finding, we discovered in this second project that people changed less because of facts or data that shifted their thinking than because compelling experiences changed their feelings. This emotional component was always present in the most successful change stories and almost always missing in the least successful. Too many people were working on the mind without paying sufficient attention to the heart."*

Much of the integration process involves modifying equipment or changes to the formal written management systems. However, all of that work will be done by and through the efforts of various individuals. As so much of the change involves people, it is vital any integration process addresses the need to shift culture and the behaviors of people. And this change must be not only in the newly acquired business but the parent company as well to be successful in achieving the expectations of senior management.

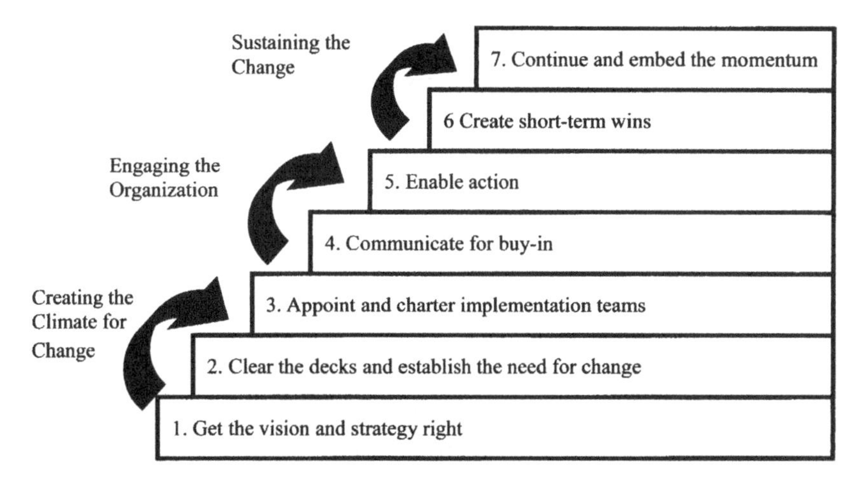

Figure 1. Kotter and Cohen's Seven Step Process for implementing Change from *The Heart of Change (2002)*

1. The first step involves creating a clear, inspiring and achievable picture of the future. The vision should address key behaviors required in the future as a necessary prerequisite to changing systems, processes, procedures, etc.
2. Secondly change leaders (i.e. those who are more than just proponents, more advocates of change) must build a sense of urgency for the needed change. To do this the change leaders will have to reduce fear, anger and any complacency that exists among staff. Further they will need to clear away any initiatives, activities, etc. that might divert or drain energy and motivation away from the primary tasks of integration which lie ahead.

3. The next step is to mobilize and organize the change leaders into assigned teams. These leaders necessarily need to be those who are focused, committed and enthusiastic about the change because those individuals:
    a. Have an understanding of the why, what and how of the change,
    b. Model the right behavior(s), and
    c. Hold both themselves and others accountable for results.
4. The change leaders and teams individually and collectively must deliver candid, concise and heartfelt messages about the change in order to create the trust, support and commitment necessary to achieve goals and targets.
5. The most challenging aspect of the next step – enabling action – is the ability of the change leaders to 'bust barriers' that hinder people who are trying to make the new vision work. Those who are developing and aligning new programs, processes, systems, etc. with the vision as well as identifying old mechanisms that are ineffective or inefficient will continually encounter roadblocks to their progress. The change leaders must be willing to step back, often from their own entrenched positions, and be able to change their own paradigms as well as those of others at their own and levels above and below them. Then these individuals will need to be decisive when committing to the new concepts and processes.
6. Achieving visible, timely and meaningful improvements thereby demonstrating progress is a vital part of continually re-energizing individuals and the new organization as a whole. Short-term wins are fundamental to keeping the process moving forward.
7. "Victory" can only be claimed after the new behaviors, processes, practices are wholly embedded into the fabric of the day-to-day operations. Only when those behaviors and practices are recognized and rewarded as the 'way we do business here' can the change leaders declare their work is complete.

Cohen and Kotter provide a few of what they term are 'key principles' to be followed when implementing a change:

- *Each step is necessary* – They note each step sets a foundation on which to build change. Omitting steps, at the least is likely to extend or impede the change process or worse possibly derail it all together.
- *The process is dynamic* – Large-scale transformations are never straightforward. Nor are they linear as depicted in the above model. It is possible to start a change process by, for example creating short term wins (i.e. step 6) as a means of initially energizing the organization thereby creating a climate for change. However Cohen and Kotter go on and stress when such an approach is adopted, it is still necessary to go back to step one, create the vision and strategy and complete all the steps.
- *Several steps can happen simultaneously and continuously* – In particular there is a need continually to communicate throughout the process. Critical to sustaining the pace of the change process is the need to communicate the urgency and energy for the required changes by all key individuals.
- *Change is an iterative process* – It is not uncommon to have to retrace steps as part of successfully moving forward. The step of creating and sustaining the urgency or energy for the change often has to be revisited as the change process progresses. Further, it is likely to prove necessary to make changes both in the composition of as well as the objectives for the implementation teams several times throughout the change process. As described in Chapter 5, plan for the need to revise teams and team members.

The above model describes a generic process for implementing change. However it does not delineate what actually might need to be changed to support the new culture, required behaviors, revised or new systems, etc. To address 'what' might need to be changed, again a variety of models exist. From those a model developed by Richard Pascale and Anthony Athos while with McKinsey's known as the McKinsey Seven 'S' model depicts the various factors a change program must address.[44]

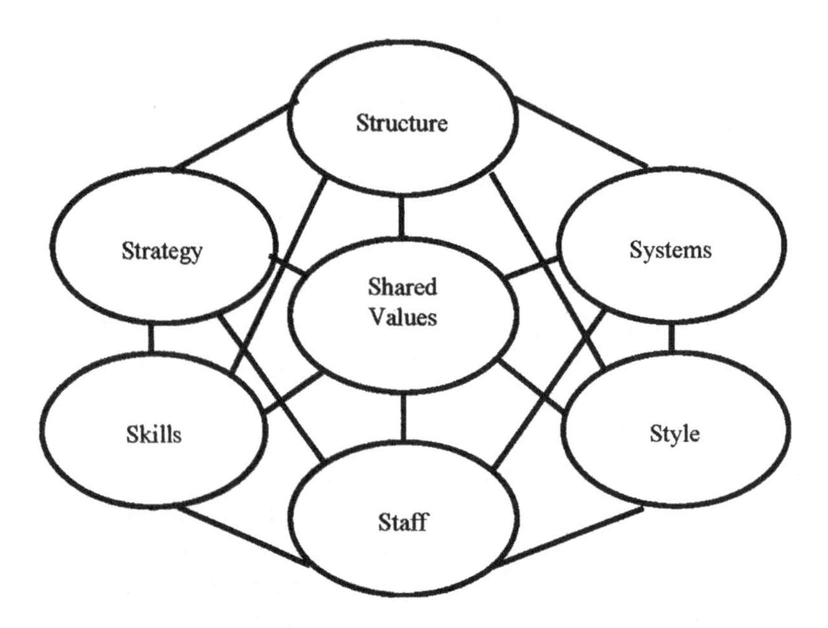

Figure 2. McKinsey's Seven S model of an organization

A short description of what comprises each of the seven "S's" is provided below:

- *Shared Values* – act as the organization's conscience, providing guidance in times of crisis (Pascale later wrote 'shared-values' is actually the same as culture but to make the seven S's work they re-termed culture, shared values).
- *Style* – refers to the employees sharing a common way of thinking and acting. They noted there are frequently several 'styles' in an organization including a leadership 'style', a functional 'style' (e.g. engineering basis, marketing driven) etc.
- *Skills* – are the skills in place to implement and deliver on the organization's strategy.

- *Strategy* – is the direction of the organization as well as the manner in which the organization and its leaders derive, articulate and communicate that direction.
- *Structure* – is the manner in which a company organizes its activities and functions. For example, whether the organization takes on a hierarchical or flat approach to decision-making and communication.
- *Systems* – includes not only the written policies and practices, it more importantly refers to all the decision making 'systems' within an organization ranging from management intuition to structured computational systems. It typically includes all computer systems, HR systems, financial and accounting systems, etc.
- *Staff* – the organization has hired able staff, trained them and assigned them to the right tasks. Key factors include selection, training, reward, recognition, retention, and where necessary replacement of staff.

The lines connecting the elements depict the interrelations that exist between the various elements. The Authors noted when one element is changed it requires the other elements be examined both individually and as a collective whole to assure a stable state is maintained. In later work, Pascale and Athos noted changes made to one element without re-examining the other elements resulted in tensions developing that eventually led to a disequilibrium causing the change process to fail. For example if developing and implementing a new accident/incident reporting system, part of the integration must be to examine and plan for how this will this impact staff, skills, style, etc. Then decide how these other elements will need to be changed to support or drive the revisions to the reporting system. Not addressing these other elements and the interactions between them will result in the new reporting system being met with resistance and skepticism.

As an organization moves through the changes it desires to make in the approach to process safety, the PS lead will have to continually assess various individual changes and their impacts on the totality of the framework to minimize tensions that might develop.

Summary – An integration process is a change process. For it to be successful you must engage the feelings or emotions of those who will be impacted by it more so than providing a series of 'logical arguments' of the need for the change.

6.2 THE INTEGRATION PATH FORWARD

6.2.1 Step 1 – Get the 'hearts' of the newly acquired business leads to accept the Vision and Strategy for the integration process.

Arrange a workshop with the executives and leaders of the overall integration process to craft and/or re-craft statements of their vision for the overall integration and a subsidiary vision and strategy for process safety issues. Make sure these speak to the hearts of individuals and engage their feelings more so than trying to rationalize the need to integrate the two businesses by just discussing the benefits. This task will not be easy as most senior managers and for, that matter, process safety professionals lean towards being rational in our arguments and discussions. Using the words of Kotter and Cohen –

'...the emotional component was almost always present in the successful changes...'.

The executive and integration leadership team need to develop a series of messages that will speak to and engage 'the emotional component' of first line managers, technical staff, etc. who are key to the process safety program. Once developed, the executive integration leadership team and the process safety integration leadership team must align their communication plans and actual communication activities. Both groups must repeatedly and continually communicate and live these messages through their own actions and behaviors,

To support this vision or strategy, simple messages of model behaviors need to be developed. Those who are key to the process safety integration effort must live those values and model the behaviors themselves before asking it of all those they interact with. It is possible these new behaviors will conflict with existing standards of behavior in the newly acquired business hence staff will question the need for such change. In these situations, rather than criticize or query staff, support and guidance must be provided to reinforce the message as well as help shape new behavioral styles. These actions are fundamental in establishing the urgency of the need for change.

Step 1 – Summary:

- *Identify change champions among key individuals that can impact process safety,*
- *Engage these champions in developing stories and communication material that engages emotions as well as logic that will present a clear vision of the 'future',*
- *Then – Communicate, Communicate, Communicate and then – communicate some more! Remember though, communication is a two way process. Without listening, just repeating the same message(s) will not achieve the required changes.*

6.2.2 Step 2 – Appointing and chartering Integration Implementation Teams

Change programs take one of two general forms, those driven from the top-down versus establishing a framework of improvement teams 'at the shop floor level' or a bottom-up approach. The 'bottom-up' approach is the concept underlying the Japanese approach typically referred to as 'Kaizen'.

The appointment and chartering of implementation teams is dependent on which approach an organization decides to take. A bottom-up approach generally entails empowering a number of teams to work on a series of specifically assigned and discrete tasks, for example 'focus teams' assigned a well-defined task. A top-down approach on the other hand means the teams will be focused more on the four major threads of an integration process (i.e. changes to the physical equipment, changes in the management systems, etc.). These teams may well establish a task force(s) under their direction to work on a specific issue, activity or procedure but they will retain the responsibility for the overall changes within a specific aspect and the interactions or implications those changes will have on other elements.

Studies of various change programs have identified two critical elements that need to be considered when deciding whether a top-down or bottom-up approach has a greater chance to succeed. Those factors are:

- The degree of consistency or agreement on the need for and the general direction the change should take both vertically and horizontally throughout the organization, and
- The degree of consistency vertically and horizontally in the organization on what needs to be changed as well as how to reach those objectives.

Where there is a high level of agreement on both the general direction and how to get there, the research indicates a bottom-up approach has greater success. On the other hand where there is a low level of agreement the research suggests a top-down approach. It is likely in most acquisitions or mergers that the degree of consistency towards the desired change as well as agreement on the need for the change will be at the lower end of the spectrum. Simply put, most process safety related integrations are likely to favor a 'top-down' approach. In such cases, implementation teams along the following lines should be appointed, chartered and empowered:

- A *guiding team* – consisting of at least one senior or executive manager and key process safety leaders who understand the need for the change, are committed to the change and have a combination of positional and personal stature and respect in and amongst the current and newly acquired business staff for them to follow the guiding team's lead.

- *An 'intolerable risks' team* – this would be an ad-hoc team mobilized to put in place temporary corrective actions for all situations or risks identified to be 'intolerable' as well as lay-out the requirements for more definitive fixes to such situations. Rather than a separate team it could be this issue of critical risks is the first order of priority assigned the physical equipment team below.
- *A physical equipment team* – this team would be responsible for identifying any changes or modifications to the physical plant and equipment, process operations, etc.
- An *organizational team* – that examines the means by which the two organizations' process safety activities are currently organized and identifies changes to the structure to assure the process safety 'organization' is aligned with the strategy and vision of the overall newly combined business.
- A *critical processes and systems team* – to map the current process safety related systems and programs identify revisions, changes, modifications and additions that will be required in those systems.
- A *people team* – to identify the competencies required and to develop training, mentoring, and recruitment programs to assure those competencies are in place as well as implementing the required shift in culture(s).

OK, now the reader is likely to sit back and note –

> "Wow – if each team has five or six members on it you are talking about assigning upwards of thirty different individuals just to the process safety integration task. That'll never fly in our organization".

Yes, you are probably correct. In anything other than a large or mega-merger, senior management are unlikely to agree resourcing several such teams. Especially in an environment where they might consider the integration process a drain on resources. In addition depending on the size of the gaps, if any, identified in the newly acquired business' approach to process safety the changes required may not necessitate mobilizing a certain team. It is also possible you could address the resource demand by phasing the mobilization of some of the teams. Depending on the various gaps found in the current state of the process safety programs, it may mean the work of some teams can be deferred until later in the integration process. It might also be the case that certain groundwork has to be laid before a particular team is mobilized. Alternatively the functions of some of the teams might be combined. For example, the functions of the organizational and people team might be combined into one team. However, whether through mobilizing the various teams in phases or combining responsibilities and functions a full-scale integration will require all of the above issues be addressed and properly resourced in some manner.

The Guiding Team

The first task of the guiding teams, as noted earlier, will be to develop a series of statements on the vision, strategy, and culture for the future approach to process safety. We again stress the need for this material to engage the hearts and emotions of the staff. The guiding team will next need to enroll the rest of the leaders in the parent and newly acquired businesses. The objective here is to enlist these leaders with modeling the future behaviors and communicating the need for change throughout the newly combined business and possibly certain outside stakeholders as well.

The overall functions of this team will include:

- Developing and communicating a shared sense of purpose for the required changes,

- Establishing clear roles for the guiding team itself as well as the other teams,
- Continually interpreting or translating the vision and strategy into goals, targets, objectives, etc. that are more easily understood and considered achievable for the other teams and the business as a whole,
- Ensuring that each of the other integration implementation teams has established effective processes for meetings, planning, problem solving, decision making, and conflict resolution,
- Creating strong relationships within and among the guiding team members and between the guiding team and the other integration teams as well as between each of the other integration teams,
- Assuring there are effective interfaces between the management structure and the various implementation teams and actively 'busting barriers' when brought to their attention by the other teams,
- Continually reviewing the progress of the integration process as well as its direction and testing whether that direction needs to be modified in light of changes in the external business environment within which the newly combined business operates.

The executives and leaders assigned to this guiding team should plan to devote somewhere between ten to twenty per-cent of their time to the change process activities as described above.

The 'Intolerable' Risks Team

Where the other teams described in this section are envisioned to be standing committees or teams chartered for the length of the integration process, the intolerable risks team is more likely to serve on an ad-hoc basis. The focus of this team will be to investigate and evaluate identified activities, processes, systems, equipment, etc.

where prompt attention and more importantly immediate or short-term corrective actions are required. Further it is likely as the integration process progresses and more is learned of the newly acquired business operations critical risks may be uncovered. As a result this intolerable risks team may be commissioned and mobilized not at the beginning of the integration process but at later stages.

The possible functions for this team may entail:

- Investigating certain situations or state of physical plant and equipment, process or operations, control systems, emergency plans and equipment, past process safety studies, etc. to further clarify the current state and whether the associated risks at intolerable levels or not,
- Develop temporary controls that can be implemented within thirty to sixty days to reduce the risks ideally to a tolerable level,
- Make recommendation(s) on additional temporary controls that might require engineering or administrative changes and would take longer than ninety to one-hundred twenty days to implement,
- Make recommendations on permanent controls that are in line with the parent company's engineering standards or require significant CAPEX or OPEX funding or resource allocations to implement.

The composition of this team will reflect the issues or risks identified as intolerable. In general though, the team is likely to have more of a focus on the physical plant and equipment than the formal management systems that comprise a sound process safety program. However, situations could well be identified requiring a focused program of training, shifts in behaviors, etc. either in parallel to or as the primary means to address certain critical risks.

The Physical Equipment Team

This team will be focused on ensuring:

- That the state of the current plant and equipment is fit for purpose, and in accordance with Recognized and Generally Accepted Good Engineering Practice,
- The design limits for the plant and equipment are known and critical operating limits and controls keep the actual operations within the design envelope,
- Whether the design is in accordance with the parent company's engineering standards.

The possible functions for this team may include:

- Reviewing past studies, engineering drawings, maintenance, test and inspection records to assess the state of these and ultimately the known state of the physical plant and equipment itself,
- Where gaps in this material or information exist, commissioning whatever required studies, inspections or tests are required to ascertain the state of the physical plant and equipment,
- Where it is considered the physical plant is not fit for its intended service, arranging for the necessary maintenance or repairs to bring that equipment into compliance.

This team may also be tasked with identifying and making changes to the current mechanical or asset integrity management systems to ensure the physical plant continues to be maintained in a manner that is 'fit for service or purpose'. Alternatively that task might be assigned to the systems team as described later in this section.

The physical equipment team, more so than any other team, is likely to face the challenge of handing senior management an itemized list of expenditures likely to cause 'sticker shock'. To prevent or minimize immediate negative reaction the physical equipment team need to be prepared for this eventuality by assuring their recommendations are well substantiated.

The Organizational Team

Kotter and Cohen note in their work, two of the four main barriers that in their words

'...often need busting!'

are organizational structures and systems. Cohen and Kotter observed that the organizational structure often fragmented a change effort. This is especially true where various initiatives require people from multiple groups to work together but those resources and the authority to implement the required changes are 'fragmented' through-out the organization. Where this is identified, the guiding team must take immediate actions and decisions to forestall such conflicts (i.e. busting potential barriers). The organizational team on the other hand will take a longer term view, analyze the root causes for such conflicts and recommend mid to long term corrective actions to the systems and/or structure.

The overall purpose of the organizational team will be two-fold:

1. Examine the current organizational structures, seek out 'synergies' that senior management will be looking for in the manner by which process safety issues are managed, supported, etc. Then evaluate whether the structure effectively and efficiently provides the proper authority to various management teams and individual managers to deliver the full range of services,
2. Develop a new structure that reflects the objectives and strategy as laid out by the guiding team.

Potential activities that might be assigned to the organizational team include:

- Analyze the current activities (i.e. actual work steps) or functions and identify which critically support the delivery of the process safety related services and map these to the organizational structure.
    - Determine the interrelationship between such activities and whether the current structure allows for efficient and effective communication, decision-making and completion of these activities.
- Analyze, in liaison with the physical equipment, critical systems and people teams those activities and the work steps required to deliver process safety services.
    - Compare the activities of the critically required services with current services. Identify common activities or work steps and/or new and unique steps.
    - Determine the interrelationships between the required activities and their potential needs or impacts on communication, decision-making and completion of those activities.
    - Evaluate where and how the activities and/or work steps should be grouped for effective and efficient delivery of those services.
    - Map the grouped work activities to the current structure.
    - Identify inefficiencies in the current structures and propose revisions to the structure to address those inefficiencies.
    - Liaise with the guiding team and develop a plan to implement changes.
    - Implement changes.

- Liaise with the systems team to identify where conflicts are arising or are likely to arise between the current process safety systems and the organizational structure. Develop mid to long-term solutions to such conflicts.

- o Lay the groundwork for changes to the systems and/or structure,
- o In liaison with the guiding and system teams implement these changes

The challenge to any organizational team is that of dealing with 'sacred cows' of functional and individual silos that become established or even worse entrenched over time and are no longer aligned with the desired strategy laid out for process safety.

The Processes and Systems team

While the structure of an organization facilitates the vertical path(s) of communication and decision making, it is the systems in an organization that facilitate 'horizontal' paths of communication and decision-making. As an example, the time and expense reporting systems of any company cut across the organizational boundaries of almost all organizations. They layout the requirements for such activities no matter what 'box' a particular individual occupies on the organization chart. An optimum is achieved when the organizational structure and its formal systems complement each other. A feat rarely achieved but worth striving for.

The second barrier frequently fragmenting a company and a change process are its organizational systems. As with structure, specific organizational systems and processes such as the information technology, performance and talent management, and training systems will perpetuate behaviors that may well hamper the progress of change. Koontz and O'Donnell note in their work that systems almost always follow a principle of "Unity of Objective".[45] In other words the various systems are designed to achieve a defined objective for that singular system. When developed such systems rarely consider the second principle of all systems namely, they are always part of a still larger system. As a result, the individual system may be one hundred percent efficient and effective delivering its singular objective but highly inefficient when aggregated into the larger whole of delivering on the organization's overall strategy. Further, once a system is in place, the resources and energy required to modify it, to achieve new goals or objectives is often substantial. A default position will develop and take the form of, revising everything else around that system is 'better' than revising the system itself.

When establishing possible new systems for process safety, remember those systems will impact or interface with numerous other systems that are also likely to be in a state of flux as the full integration of the acquired business takes place.

The overall purpose of the process and systems team will be to develop an understanding of the capabilities and limitations of the current systems and processes and then assess those capabilities against the expectations which have been established for the new approach to process safety that is to be implemented. Of import is determining when the limitations of the current system(s) require major changes, modifications, additions or the possible replacement of current systems.

Potential activities that might be assigned the process and systems team include:

- Map the existing support processes and systems that directly support or enable the efficient delivery of process safety requirements,
    - Review how applicable the current support mechanisms are to the new approach or services established for process safety,
    - Determine if there is a need for new specific dedicated functions to support the new services,
    - Build on existing systems to integrate common processes and systems,

    - Develop any required changes and, in liaison with the organization and people teams, implement those changes.
- Describe the current and future flow of process safety information within the organization as well as to customers or other concerned stakeholders (e.g. regulators, NGO's, local authorities, etc.).
    - Map current work processes to identify and understand current information flows and the need for any changes in accordance with the PS strategy,
        - A maxim – build on existing work processes and systems as much as possible.
    - Assess the current knowledge infrastructure, develop mechanisms to capture and retain critical information and to input this knowledge into the organization's IT systems,
    - Develop programs to systematically gather and codify knowledge of a high quality,
    - Identify PS knowledge islands and ensure this knowledge is being properly shared or disseminated throughout the organization,
    - Develop mechanism(s) to provide this valuable insight to key line managers, supervisors and staff as needed,
    - Clarify interfaces with other interested stakeholders or groups,
    - Design processes for linking up to and utilizing external knowledge sources with regulatory bodies or appropriate industry associations.
- Develop necessary supporting mechanisms or processes.
    - Modify incentive systems adding KPI's and performance measures to ensure staff contribute to and use the knowledge management systems,
    - Design the processes to be innovative and support future improvements in process safety.

The People team

The overall purpose of this team will be to manage the people transition process in a way that makes staff excited about the potential opportunities that will accompany these changes.

Developing the necessary tools, processes, services and practices, to achieve a goal of a process safety program that accords with good practice may mean the people skills will need to change as well.

The people team will need to define explicitly the new behaviors and identify the skills, abilities, and attitudes required for the staff to succeed in a new working environment for the business as a whole as well as the process safety aspects.

Potential activities that might be assigned to this team include:

- In conjunction with the organizational and systems teams, identify the skills and capabilities required to deliver required process safety related services.
    - Develop training and/or recruiting programs to assure these skills are promoted and assimilated in a timely manner.
    - Test how the new skills are integrating with current skills or services,
    - Identify conflicts and develop programs to address any conflicts.
- Develop programs to imbed skills (e.g. social skills) required to make a successful transition.
    - Institute these programs and test their effectiveness.
    - Identify deviations to desired changes and develop solutions.

- Assess and enhance the culture of the line management, technical support teams, supervisors, operators and maintenance staff towards process safety;
    - Ensure the behaviors, mores, and values of each group are in line with the desired attributes of a strong approach to process safety,
    - Identify gaps,
    - In liaison with the guiding and organizational teams, develop and begin programs to shift culture(s) where needed.
- Examine and modify performance appraisals to include measures that demonstrate commitment to the process safety vision, strategy and change process.
    - Develop a process or mechanism to address behavior(s) that is resistant to the change.
- Ensure succession plans promote individuals who support the process safety vision, strategy and change process and will serve as role models to others.
- Ensure the recruiting and hiring systems select individuals who have the skills as required by the new needs.

Much of the people team's work, will follow-on from the activities and findings of the organizational and systems teams. As a result, the work of the people team could possibly be deferred for a period of time. However it is important to have in place mechanisms to reward positive behaviors and correct negative or resistant behaviors as implementation progresses. The guiding team or process safety integration lead should balance the possible resourcing issues associated with mobilizing the people team later against the benefits of appointing and chartering the people team at the same time as the other teams.

Step 2 – Summary, appoint and charter implementation teams of the following nature:

- *An 'Intolerable risks' team, to identify short term corrective actions addressing situations or risks the company would deem are at an intolerable level*

- *A Guiding team, consisting of at least one executive or senior manager with various responsibilities, one being the authority to 'bust barriers' that become an impediment to the integration process moving forward,*
- *An Organizational team, with the objective of identifying synergies in and among the current process safety related structures as well as assuring any future organizational structure delivers an effective and efficient process safety program,*
- *A Physical Equipment and operational process team to identify gaps in the state of the current equipment and/or process and develop cost effective solutions to such variances.*
- *A Process and Systems team to identify and implement the critical process and systems, including IT systems, to support and help drive the desired approach to process safety,*
- *A People team to assure the necessary skills, competencies and culture exists to sustain an effective process safety program.*

6.3 AN ALTERNATE BOTTOM-UP APPROACH TO INTEGRATION

The above discusses the first step in taking a 'top-down' approach for integrating or merging two approaches together. The following discusses the form a 'bottom-up' approach might take.

Rather than appointing a series of standing committees (i.e. a guiding team, a systems team, etc.) a bottom-up approach is more likely to appoint a number of focus groups or task forces on an ad-hoc basis. Each focus group would be assigned a specific task such as writing or re-writing certain manuals or procedures, integrating the engineering standards or requirements, merging incident and accident reporting systems, etc. The focus of such teams would be to close specific gaps or a collection of like or related gaps found in the current approach to process safety. The tasks or gaps these focus teams are assigned should be prioritized taking into consideration the risks associated with any identified gaps, the ease or ability to address a gap, the availability of resources, etc. Going back to Kotter's work, to build momentum for the integration or merger process as a whole, you might find it of value to mobilize a number of teams to focus on possible short term wins (Step 6 in Figure 1). This will help overcome the inherent inertia both in the newly acquired business or facility as well as the parent company towards change. It may also help in creating a small corps of process safety change agents, advocates for additional and further change in the process safety programs, systems, approach, etc.

The tasks assigned focus groups should be 'short and sharp'. Ideally a focus team should not be assigned a task that extends beyond a period of six months. Where it is recognized a particular issue will take longer to change, it should be broken up into steps each projected to last no longer than six months. The options always exist to re-charter the existing focus team(s) to work on the next step of the change possibly supplementing or replacing members versus chartering a whole new team.

Each focus team mobilized should be formally chartered. The charter should contain a clear explanation of the current state, the desired outcome including a tangible deliverable and a target date for the deliverable. A proposed working process for the focus team might be included in the charter. However, often as a means for the helping the proposed members themselves to form into a team, the approach is taken for the team to develop an outline of their proposed working process and provide that within a period of a week or so to the guiding team for agreement. Where a focus team will need to interface with other focus teams, a possible mechanism for such liaison should be included in the charter.

It is likely a number of such focus groups will be active at any-one time throughout the integration process. There is need then to oversee their individual and collective progress. Further it is likely these focus teams will have limited authority to address organizational related issues, authorize and mobilize external resources or contractors, etc. As a result, as with the top-down approach, there will be a need for a team something similar to the Guiding team. Even with a guiding team in place when a number of focus teams are active, you might consider appointing a full-time coordinator to act as a single point of contact between the guiding and various focus teams. The role and authority of the coordinator could extend from that of just being a channel for communications to having the power to call on other resources within the organization, bring in outside expertise where required, etc.

When a focus team completes a task, there is a need to 'advertise' their success both to the other focus teams as well as to the organization as a whole. This will help build or sustain the momentum among the teams and the organization a whole towards an eventual successful integration. Conversely where a focus team finds it is not possible to complete their task either within the timeframe as originally agreed or where in the opinion of the focus team it is unlikely the required changes can be made at all, this fact too needs to be communicated. The reasons for any delays or the possibility a change cannot be made should be discussed transparently and any proposed alternate solution provided.

Adopting a top-down versus a bottom-up strategy of merging two approaches need not be mutually exclusive. In fact, it is highly likely that in a top-down approach situations will arise where, for example, the team responsible for systems identifies a particular set of activities that need to be accomplished. The systems team may well mobilize a focus team to address that particular issue. It is also likely that in a bottom-up approach a whole series of individual activities will be identified. On examining what might be required to complete those activities the need to convene a broader mandated capital improvements team (i.e. a physical equipment team) might be identified.

Step 2a Summary – using a bottom-up approach

- *Appoint a team to coordinate, oversee and address issues beyond the scope of the various focus teams' authority,*
- *Prioritize the various activities and tasks to be addressed,*
- *Formally charter each focus group, and assure the charter contains a deliverable as well as a target deliverable date.*
- *Keep tasks 'short' ideally no longer than six months and break large changes into shorter tasks where necessary,*
- *Widely communicate the success(es) of the focus teams to build and sustain momentum for the changes.*

6.4 DIFFERENCES BETWEEN FACILITIES, BUSINESS UNITS

To this point, the discussions have approached the integration phase of the newly acquired business(es) as a whole or uniform group. However it is probable there will be significant differences by which process safety is perceived and approached between operating divisions, individual facilities, geographical areas or locations, etc. of the newly acquired business' operations.

When integrating together the various facilities that comprise a newly acquired business it is vital to recognize these differences. Where new process safety programs and approaches are developed they will have to be tailored to the general needs of the various individual facilities, business units and possibly even at the strategic business unit level. As an example, a 'full-fledged' process safety or major hazard accident prevention program requires the duty holder to have a sophisticated Integrity Management program in place. Such a program will incorporate the principles of an integrated Safety, Health and Environment program. Only larger organizations or facilities will have the resources to be able to develop, implement and sustain such programs. To require such an approach at smaller sites where the risks are different would be at odds with one of the attributes of leading edge practice – namely that policies and programs should be proportionate to the risks they are controlling.

When moving forward with implementing a new approach to process safety an element of 'proportionality' should be adopted. As a result it may be necessary for the new approach to adopt a 'parallel universe' concept to process safety oversight. For the larger facilities and business units the requirements might well be that of a best practice nature. In smaller facilities the approach adopted might be one less resource intensive yet still contain the ability to ensure the risks of a major hazard or process safety event are controlled to tolerable levels. As a result, the process safety implementation team(s) may have to develop and be capable of delivering a process safety service package that meets what might at times be viewed as disparate needs.

6.5 STEP 3 - WORKING THROUGH THE IMPLEMENTATION ITSELF

You have the gaps between the current and desired states. You have your plan. You have decided to take a top-down or bottom-up approach and have identified and mobilized various teams. The only thing left is to start the actual emotional and physical grind of working through all the various steps as laid out in your plan.

The Process Safety integration lead will find they play many roles during actual implementation. They will be a focal point of communication between the various teams and the assigned senior management, or executive sponsor of the whole integration process. They have the role of project managing the totality of the work being done by various groups or teams. A caution here is that to the best of their ability the PS integration lead should try to avoid project managing the activities of any one particular team. That should clearly remain the responsibility of each individual team.

The PS integration lead is also likely to be viewed as a focal point between the existing line management and the integration change teams, etc. These roles will change as the integration or merger process progresses. As a result, it is easy for the process safety integration lead's energy, time and individual capabilities to be totally consumed by such activities. Where the integration process has been assigned as an additional responsibility to the lead's regular duties or activities, they will be constantly walking the tightrope of how much time they have to spend on integration related activities versus their 'normal day-to-day' job.

While the varying demands will place considerable strain on the PS lead personally, it can also stress and strain the potential success of the process safety integration as a whole.

There are three issues fundamental to a successful change process.

- One is developing and constantly reinforcing a clear vision of the future.
- The second is that the leadership and individuals who are key to success must actively model the behaviors of the future.
- The third is – communicate, communicate and then, communicate some more.

As noted earlier it will be easy for the process safety integration lead's time and efforts to be consumed in mainly mechanical tasks. Where this occurs, unless a senior manager or executive sponsor has from personal experience developed their own individual passion for process safety, it is likely these three issues will not be given the priority they require. When this happens the chances of achieving a truly successful integration will drop significantly. The process safety integration lead must build time into their own personal integration related activities and responsibilities and ensure:

- The vision for the future approach and program for process safety is clear and compelling,
- Executive, senior and line managers as well as key individuals model or emulate this vision in their own activities and in carrying-out their own duties or responsibilities,

- Compelling stories of the need for the future approach are developed and the progress towards achieving that future is constantly communicated.

Mechanical changes to the physical equipment or processes, revisions and modifications to the written or formal systems can and will occur. However, these changes must be accompanied by shifts in the culture and behaviors of key line managers, supervisory as well as operating and maintenance staff. Unless such shifts occur, the changes made in the formal systems will at best be a pyrrhic victory.

Whether taking a top-down or bottom-up approach, make sure the individual teams as well as the integration process as a whole includes time, resources and activities to embed the changes into the new normal operating activities. In major engineering projects, time and resources for integrating the new facilities into the operations of the existing plant and equipment are always planned for at the beginning of the project. However, it is not unusual that budgeted allocations for 'integration' are eroded away when the project runs into delays and overruns. The same also often happens in an integration or merger. Considerable time and effort might be allocated to embed the changes into actual operations at the beginning. However when it actually comes time to begin embedding the changes into the day-to-day operations, this often becomes no more than an announcement on the intranet that a new set of procedures are now available and time should be set aside to read them. Where this happens it is highly doubtful the new procedures will ever be read let alone implemented versus being truly embedded. Each of the teams must make provision in their activities to embed any changes. In addition, the process safety integration lead must continually seek the commitment and support of senior and line managers to ensure the necessary time and resources will be set aside for training and changing behaviors of staff where and when required.

To confirm the new changes have been effected, the group or team responsible for operationally auditing organizational activities should be provided with material detailing the changes made which they can incorporate into their audit manuals and protocols. Discussions should be held with this group to determine whether a special audit of all the process related safety changes should be undertaken after a period of a year or at most possibly two years. The focus of such an audit would be on identifying how well the changes are being effected in the normal day-to-day operations. Alternatively the operational audit group might be of the opinion it is better to address such changes as part of the normal planned operational audit schedule. However, it may be that certain changes are considered critical to controlling high-risk situations or scenarios. In these cases, agreement should be sought from senior management of a need to ensure the required changes are in fact being followed and are well embedded into day-to-day activities. In these cases there may be a need to undertake a focused special audit. Alternatively, this might be addressed by requiring the line management to report periodically on the status of these new controls.

Do not underestimate the amount of time it will take to agree on a new set of classifications in areas such as your document hierarchy. The emotive factors associated with decisions and agreement to change 'procedures' to 'guidance documents' or 'guidelines' are always underestimated. Many will see this as just mere 'word-smithing' but they will often have significance at the operating and craft level far from being viewed as 'word-smithing'. Where such changes are contemplated, the integration process needs to build in time and activities to introduce these changes to the workforce and be clear on what is intended by the change. Perhaps more importantly it is necessary to clarify what is not intended by the change. As an example, where it is decided to revise and rename a whole series of 'operating procedures' to 'operating guidelines' is it meant in making such a change to provide operators and supervisory staff greater latitude in using their own discretion and judgment when certain situations occur? If that is not intended, staff must be clear on the matter. It is also imperative to recognize the impact such changes will have on the current culture and the future desired culture.

Step 3 – Summary:

- *Assure the 'vision' of the desired state is clear and key individuals begin the process of modeling the new desired behaviors,*
- *As the process safety integration lead be prepared to play a variety of roles but balance these against assuring the vision and strategy of the future continues to be communicated in clear simple emotive messages by key individuals,*
- *Assure as the various teams work through their assignments sufficient time and resources are allocated to transitioning the changes to operations, do not allow the effort to achieve transitioning to be eroded,*
- *Make provision for audits of the planned changes to be undertaken after an agreed period of time such as a year or possibly at most two years.*

You are now at a point, where the next steps are in 'your court'. It is our hope this Guideline has helped with:

- Grasping the potential breadth of issues that need to be considered and addressed in the acquisition and integration of facilities or businesses that process, manufacture or store materials that are highly energetic and in quantities such that they would fall under the jurisdiction of process safety or major hazard legislation, and
- Providing assistance with outlining the various steps, tasks or activities you are likely to face when acquiring and integrating or merging such facilities into your business.

You should now be in a position to put this Guideline aside and start working through the various stages of an acquisition or merger with an increased level of confidence than when you picked it up almost two hundred pages ago.

7

M&A IN THE FUTURE

Trying to predict the future is like trying to drive down a country road at night with no lights while looking out the back window.
Peter Drucker

7.0 COURTNEY'S STORY - CONTINUED

The purchase of White Hot Chemicals had occurred a little over four years ago. Skip, the CEO at the time, retired about two years after the purchase to great acclaim at having fundamentally shifted the whole focus of Bland Petroleum through that acquisition. The purchase and subsequent integration of White Hot's operations into Bland had such a marked change they were actively considering rebranding the company from Bland to White Hot.

Courtney's efforts on the integration of various HSE issues and to a very large extent the process safety and major hazards programs did not go unnoticed. Six months ago she was promoted to the position of Vice President of Compliance and Regulatory Affairs when Carla her former boss took over as head of one of the newly formed strategic business units.

Gareth had taken over as President and COO of the company after Robert was elected to the position of CEO when Skip retired. It was the call from Gareth that caused Courtney to reflect on all these changes. At the time it was exhausting but looking back at it, it was also very exciting.

When she entered Gareth's office six of the eight various Vice Presidents and the two regional Presidents were there. Courtney was a bit surprised as she wasn't aware of any scheduled executive team meetings for another six weeks. Courtney's first thoughts were here we go again another acquisition!

Gareth asked them all to be seated at his conference table. He immediately took the lead and told them approximately two weeks ago a joint venture of an Indian and Chinese company had approached Bland with an offer to buy Bland that was most attractive. The full Board was being convened that afternoon and it was likely the recommendation to the Board would be to accept the offer!

Now in this Industry what kind of an ending did you expect?!?

As we write this Guideline in early 2009, M&A activity worldwide has slowed to a snail's pace. The ability to finance possible acquisitions, as of early 2009, underlies this slow down and is the direct result of a worldwide credit crunch. That said Dow Chemical did close on the acquisition of Rohm-Haas in April 2009. A deal estimated at US$16.3 billion. So there is still 'some life' in the world of M&A. Eventually the current worldwide economic slowdown will end. When that occurs the strongest surviving companies will accelerate their M&A activities since any company that does survive the 2008 - 2009 economic turmoil will be an obvious good company to consider.

As companies emerge out of the economic turmoil of late 2008 and early 2009, what trends might those in the petro-chemical or broader based process industry follow when considering a future acquisition or partner to merge with? At the expense of being accused not only of having switched off our lights and turned in our seat to look out the back window in offering the following, you might also think we donned a set blinders, as well – *but here goes*:

- At a Stanford University 2006 economic symposium, some of the economists put forward a view that state based National Oil Companies (e.g. Saudi-Aramco, Kuwait Petroleum, Petrobras, Petronas, CNOP, etc.) will over the next few decades become the dominant players in much of the world's petrochemical industry. Some even predicted many of today's international or independent oil companies (e.g. the Conoco-Phillips', the Valero's, etc.) would find their core business transformed into being service providers for the NOC's.
- In the February 2009 Harvard Business Review it noted, nine of the top ten wealth funds are state or government owned or backed.[46] Leading this list is the Abu-Dhabi Investment Authority with almost US$800 Billion of assets. Second in the list is the Norwegian Government's Pension Fund. The combined assets of these two funds were estimated at US$1.2 Trillion. However, while impressive this is still only about 10% of the United States debt, which is projected at US$11 Trillion. In contrast, the total combined assets of more familiar private equity firms or names such as - The Blackstone Group, The Carlyle Group, Bain Capital, and Kohlberg, Kravis and Roberts were estimated to amount to around 10-15% of the assets of the Abu Dhabi fund alone. Where an M&A activity faces or is being driven entirely by financial issues this can result in the resources required to implement adequate process safety programs being reduced to non-sustainable levels.
- In a May 2009 article, Nirmalya Kumar (Professor, London Business School) noted how the "emerging giants" from India, China, Malaysia and Russia were rewriting the typical

rules regarding Mergers and Acquisition.[47] One such rule Professor Kumar contends that is being re-written was unlike Western companies, which use M&A primarily to increase size and efficiency, these emerging giants were acquiring to obtain competencies, technology and knowledge.

- Just prior to the 2008-2009 economic downturn, Jack Welch (famed ex-CEO of GE) noted that in the next decade productivity increases of 6%, which were considered 'world class' in the past would be viewed as 'barely keeping up'. He predicted companies would have to be turning in 8-9% increases in annual productivity for a company to be considered a world leader.
- 'Access to market(s)' which is often a primary driver for companies acquiring or merging with others is also viewed as continuing to be a driver, with the shift though of gaining access to markets in China, India, S.E. Asia more so than North America or Europe. Conversely, according to Professor Kumar, emerging giants in China, Russia and India will be seeking access to markets in North America and Europe. Any M&A activity that involves companies from different countries can create significant barriers to successful integration. Language and local culture are two of the most significant barriers that need to be overcome in addition to the barriers related to process safety language (i.e. local regulations and industry standards) and culture, e.g. a rules based versus risk based approach to process safety.
- Finally there is an emerging 'thread' of a view, that evolving out of the 2008-09 economic turmoil will be a trend back to smaller more specialized or 'boutique' operations. This occurred in the 1980's and 1990's in the steel industry with a move from large companies and steel processing facilities to specialized 'mini-mills'. Should this occur we would likely see more of the 'divestment' side of the M&A coin versus the 'acquisition' side.

While the above are drivers to actual acquisitions, mergers or divestment activity within industry in general, they are only one set of factors or attributes impacting the overall profile of future process safety or major hazard safety programs. The future forms or states of process safety programs are likely to be impacted more by changes in technology, demographics, socio-political moves and perhaps most important of all, shifts in norms, mores and cultures at the regional, country and local levels, etc. A few examples or 'forecasts' are provided below, more to start your own lateral thinking on how these factors could impact the future approach to process safety than to try and portend what the future might actually hold:

- Technology
  - There is little doubt the move to more and more process automation will continue. This can have a positive impact on reducing potential accidents or incidents where human error is identified as a major contributing factor. However as seen in some recent notable aircraft near-misses attributed to software errors and in others multiple software packages 'fighting with each other', totally automating a process is not without its own set of hazards.
  - Focusing on some of the activities involved in an M&A, the advances in technology and information systems means there is likely to be fewer and fewer acquisitions where the data room is a physical location populated with hard copies of documents. In addition, meetings, site interviews and even site tours could well take the form of Web-ex conferences and 'virtual' site tours rather than physical meetings and physical walkthroughs.
- Human Resources
  - As the 'baby-boomer" generation is replaced by the Gen-X generation, rapidly followed by the Gen-Y's and within the next decade the so-called 'Millennial' generation we will see totally new paradigms develop in work practices, work habits and cultures. Who knows we may even eventually get to the so-called

euphoric state of a truly 'paperless office' first discussed in the 1980's.

- Quality of Life issues will continue to escalate both in respect to the quality of the actual work-site as well as the demand for products and services that assist with living in an ever more technically complex environment.

- Socio-Politic moves and mores
    - Tightly tied with Quality of Life matters will be pressures on governments and corporations to act in more and more responsible ways. Corporate sustainability and corporate social responsibility are likely to be issues that continue to be major deciding factors in the platforms of various political parties. What will be interesting in the next several years will be the emphasis governments place on enforcing those statutes and regulations currently in place versus passing further legislation. The indications within the U.S. are that the various regulators have been told to take a more active posture in this regard.
    - The acceptance of process safety related accidents, especially where the consequences are multiple fatalities, damage to the environment or business interruptions of a scale that would threaten the on-going viability of a business will continue to decline. In the US, regulatory emphasis on process safety has increased significantly with the launch of OSHA's Petroleum Refinery Process Safety Management National Emphasis Program (NEP) on June 7, 2007. This program identified significant process safety issues during their 2008 inspections resulting in over 200 citations for the top seven PSM elements and sub elements. Further, as witnessed in China, where accidents that resulted in multiple workers being fatally injured have occurred, the repercussions on the management of those facilities can literally be life-threatening.

Do we see process safety or major hazard safety management continuing to evolve and improve? Most definitely, yes. It may well be over the next few years depending on how the current 2008/2009 economic turmoil plays-out we will see a plateauing in interest with respect to new regulations or enforcement actions. However, once we move through this period of economic turmoil, a period which is likely to coincide with some major demographic shifts as well, it is likely the professionals in the process safety field will have to undertake some fundamental rethinks of what is currently accepted practice. If we are correct in such projections and should the M&A activity 'heat-up' at the same time, the PS profession as a whole and especially those directly involved in merging and integrating what could be diverse programs together are likely to find their lives very-very exciting.

The vision of CCPS is that all companies involved in M&A activities will learn from the past and use information as contained in this Guideline and other CCPS material, to improve their evaluation of process safety related issues. In addition to the methods in this Guideline for evaluating process safety at each stage of the M&A process, going forward, companies will have a set of process safety metrics to evaluate more quantitatively the performance of companies involved in M&A activity. In fact, poor process safety metrics may become a significant deal breaker when screening potential candidates for M&A as their importance becomes better recognized. These metrics are described in detail in a recently released CCPS publication.[48] Just as the need to address environmental issues is now standard practice when considering a potential M&A, an evaluation of process safety concerns could become standard practice as well and not a 'bolt-on' as companies use these tools and recognize their value. We do not envision that the overall M&A process will change significantly in the near future. How it is performed though, is likely to be impacted by various changes in technology, training, mores, etc. Further, we believe process safety will become a "core" issue to evaluate during M&A

activities and that process safety metrics will become a major factor in the decision making process

Having dented our rear bumper on a few rocks, trees and walls we didn't see as we were backing down that good ole country road (but hopefully not having run-over anyone), it is time to turn-around in our seats, turn-on the headlights and get to work on that divestment, acquisition or merger currently staring you in the face!

THE APPENDICES

APPENDIX A – M&A PROCESS SAFETY CHECKLIST

M&A P.S. CHECKLIST – COMMERCIAL EVALUATION PHASE

Information or Issue	Date	Requested Of	Date Received	Comment on Information	Impact on Deal
What type of transaction is being considered:					
Purchase, sale, lease, shares, asset, Joint Venture, entry into new market, partial site acquisition or divestiture, etc.?					
What is the approximate value of the transaction and how big of a 'problem' would kill the deal?					
In monetary terms,					
From the standpoint of reputation(s),					
From a future resource requirement perspective?					
What are the fall back strategies for each party involved if there is a problem in the transaction and what are the possible Process Safety impacts and plans to address these?					

Information or Issue	Date	Requested Of	Date Received	Comment on Information	Impact on Deal
If a new corporation may be or is formed as part of the transaction, what impact will this have on Process Safety management?					
Who is or are the current owner(s)? Who is the operator of record if different from the owner?					
Identify whether the site(s) are required to conform to any process safety related governmental regulations (e.g. PSM and/or RMP in the U.S., Seveso in Europe, etc.)?					
Do you have any contacts in these organizations? If so who and what are their roles?					
What alternative forms of transaction are being considered that could impact Process Safety concerns?					

Information or Issue	Date	Requested Of	Date Received	Comment on Information	Impact on Deal
If alternative sites are being considered, what is known of the non-Process Safety factors driving selection of one site over another?					
What factors affect the practicality of excluding/including the land beneath the asset?					
If a lender is involved in the transaction:					
What are the lender's HSE concerns?					
What are the lender's HSE guidelines and protocols that must be satisfied?					
If the identified lender is concerned about the transaction, what factors would limit dividing the loan among several lenders?					

Information or Issue	Date	Requested Of	Date Received	Comment on Information	Impact on Deal
Will the HSE and particularly the Process Safety portion of these aspects arising from this assessment affect the interest rates? What factors might be of most importance in affecting this rate and to what possible extent?					
If more than one lender is being considered, do their HSE concerns differ? If so, is there a mechanism in place to reconcile such differences?					
If the loan includes provisions for HSE oversight by the lender(s), how is the transaction affected by these provisions?					
If the World Bank, IMF, ADB or other international financing is involved review the applicable guides and incorporate into M&A process.					

Information or Issue	Date	Requested Of	Date Received	Comment on Information	Impact on Deal
If insurance is involved as a risk management option:					
What types of coverage are in place or contemplated?					
What is the financial strength and payout record of the insurer?					
What are the Process Safety aspects of the HSE requirements of the insurer?					
What are the limits of coverage?					
Have the insurers carried out any safety or fire protection reviews or audits and the status of such audits?					
Review all the parties involved in the transaction:					
What are their individual relationships with the transaction?					
What is their financial strength?					
What is their degree of risk aversion?					

Information or Issue	Date	Requested Of	Date Received	Comment on Information	Impact on Deal
How well do they understand the Process Safety aspects of the overall HSE risks?					
By virtue of the size or status of either party to the transaction, will any of the parties be subject to different Process Safety standards or scrutiny after the completion of the transaction versus before?					
Assess whether the nature of the transaction could subject the new owner to liabilities arising from properties formerly owned by the entities involved in the transaction.					
Similarly consider liabilities that could arise from claims by former customers or employees.					
Assess the potential for any off-site liabilities.					

M&A P.S. CHECKLIST – THE M&A TEAM

Issue or Activity	Responsible	Date Raised	Resolution	Date Resolved	Future Follow-up Required
Taking into consideration the location and nature of the business(es), ensure there is a sound understanding of applicable Process Safety requirements in your M&A team members whether by virtue of compliance with:					
Statutory law,					
Common law,					
Administrative regulations or standards,					
Industry norms and standards.					
Clarify boundaries between team members for all HSE-Process Safety aspects, particularly:					

Issue or Activity	Responsible	Date Raised	Resolution	Date Resolved	Future Follow-up Required
How is the assessment of potentially overlapping areas with Process Safety issues such as security, pollution prevention, etc. to be apportioned and managed between the team members?					
How is the possible or eventual divestiture or decommissioning and disposal to be assessed and modeled?					
How will HSE related real property, easement and zoning considerations be coordinated with other groups on the M&A assessment team?					

M&A P.S. CHECKLIST – DATA ROOM INFORMATION

Information or Data Item	Requested YES-NO	Date Requested	Reference or ID Number	Date Provided	Priority	Impact on M&A (e.g. Hi, Med, Low)	Summary or Comments
What are the exact locations of current and past operations?							
Mailing and shipping addresses,							
Latitude and longitude of each facility or site,							
Legal designation,							
Exact description of which properties are included,							
Any other assets that are involved in the transaction (e.g. pipelines, terminals, ports, etc.).							
What is the current and proposed designation of the facility per NAICS/SIC code?							

Information or Data Item	Requested YES-NO	Date Requested	Reference or ID Number	Date Provided	Priority	Impact on M&A (e.g. Hi, Med, Low)	Summary or Comments
Request data on:							
The size of the facility(ies) (acres/hectares of property, area of buildings, capacity of processes and vessels, etc.)							
Date(s) each facility was built, operated, major expansions, etc.							
Site and building plot plans.							
The plot plans should clearly show each of the major infrastructure, processing and/or manufacturing facilities.							
Request that the plot plans be divided using a logical process (e.g. the process or manufacturing flow) and various sections be given an identifier or short name for later reference.							

Information or Data Item	Requested YES-NO	Date Requested	Reference or ID Number	Date Provided	Priority	Impact on M&A (e.g. Hi, Med, Low)	Summary or Comments
Are the sites or facilities subject to process safety, risk management, major hazard or safety case related legislation or regulations?							
Request copies of all documents submitted to the appropriate regulatory agency and copies of all documents from that regulator.							
Request copies of all citations, improvement notices, audit findings, observations, etc. issued by the Regulator.							
Request copies of all corrective actions undertaken to address issues raised by the regulator.							

Information or Data Item	Requested YES-NO	Date Requested	Reference or ID Number	Date Provided	Priority	Impact on M&A (e.g. Hi, Med, Low)	Summary or Comments
Request block flow diagrams or ideally a Process Flow Diagram(s) with data on the steady-state flows of all major streams.							
Request a material and energy balance for each facility as a whole and all large units in each site or facility.							
Request process design or process technology information for each site or facility that is to be included in the transaction. This should include:							
General process description,							
Process chemistry,							
Maximum intended inventory,							
Safe upper and lower limits for various process parameters,							

Information or Data Item	Requested YES-NO	Date Requested	Reference or ID Number	Date Provided	Priority	Impact on M&A (e.g. Hi, Med, Low)	Summary or Comments
Evaluations of the consequences of deviations from safe limits.							
Request information on the process control configuration, including:							
Process control strategy,							
Hardware/software used for process control,							
Configuration of control room(s),							
Process control technical support.							
Request information on the capital and maintenance budget(s) for the past 'X' years.							
Are there discernible trends in these budgets?							
Is there an obvious under-spend?							

Information or Data Item	Requested YES-NO	Date Requested	Reference or ID Number	Date Provided	Priority	Impact on M&A (e.g. Hi, Med, Low)	Summary or Comments
Request a list of major equipment with age and general condition with reference to maintenance schedules. Major equipment would typically include:							
Compressors,							
Large pumps (e.g. >50Hp),							
Pressure vessels,							
Reactors,							
Boilers,							
Heat exchangers,							
Storage tanks,							
Fired heaters, etc.							
Request a list of all identified critical equipment with, at a minimum, any legally required information, for example:							
Materials of construction,							
Engineering drawings (i.e. P&ID's),							

Information or Data Item	Requested YES-NO	Date Requested	Reference or ID Number	Date Provided	Priority	Impact on M&A (e.g. Hi, Med, Low)	Summary or Comments
Electrical classification,							
Relief system design and design basis,							
Design codes and standards used,							
Safety interlocks.							
Request a list of chemicals or materials handled or processed and related process safety information, including:							
Physiochemical and toxicological properties,							
Flammability,							
Stability,							
Reactivity,							
Corrosivity.							
Request information on the integrity and maintenance of process safety systems, including:							

Information or Data Item	Requested YES-NO	Date Requested	Reference or ID Number	Date Provided	Priority	Impact on M&A (e.g. Hi, Med, Low)	Summary or Comments
Electrical distribution preventive maintenance program(s),							
Mechanical integrity program(s),							
A selected sample of integrity test and inspection records and inspector qualifications.							
Request process safety related metrics, including:							
Leaks, breaks and spills data,							
List of incidents in the last 'X' years reported to government authorities,							
Key process safety issues or challenges,							
Any pending issues that could lead to litigation.							
Request process safety hazard assessments, including:							

Information or Data Item	Requested YES-NO	Date Requested	Reference or ID Number	Date Provided	Priority	Impact on M&A (e.g. Hi, Med, Low)	Summary or Comments
Process Hazard Analysis and control studies,							
Worst case scenarios and lines of defense,							
Facility siting analysis for permanent and temporary buildings,							
Tests of the flammability or explosivity of any dusts generated by the process or operations.							
Design of key process safety control and prevention measures, for example:							
Spacing, drainage, fire detection and protection, etc.							
Hazardous inventories,							
Layer of protection analyses,							

Information or Data Item	Requested YES-NO	Date Requested	Reference or ID Number	Date Provided	Priority	Impact on M&A (e.g. Hi, Med, Low)	Summary or Comments
Quantitative risk assessments,							
Offsite consequence analyses,							
Action items arising from any of the above studies and the status on critical items,							
Emergency or contingency plans and status of local authority and community involvement.							
Request security vulnerability assessment and site security plan.							
Request the status and follow-up of any findings arising from these studies.							
Has a budget been allocated to correct all open items?							
Request a selected sample of standard operating procedures or programs,							

Information or Data Item	Requested YES-NO	Date Requested	Reference or ID Number	Date Provided	Priority	Impact on M&A (e.g. Hi, Med, Low)	Summary or Comments
including procedures for:							
Initial start-up,							
Normal operations,							
Temporary operations,							
Emergency operations and emergency shut-down,							
Operating limits,							
Operating procedures accessibility,							
Program(s) for review and certification of operating procedures.							
Request an organization chart for each site or facility.							
Where assets are geographically distributed (e.g. pipeline operations) request information on location of control centers and where							

Information or Data Item	Requested YES-NO	Date Requested	Reference or ID Number	Date Provided	Priority	Impact on M&A (e.g. Hi, Med, Low)	Summary or Comments
operating staff and key support are physically located.							
Request the loss prevention reports from the property insurance carrier.							
Request information on all planned major projects or modifications.							
Request information on whether the associated process safety liabilities and risks of these planned changes have been assessed using recognized procedures or methods.							
Request information on how the liabilities and risks are to be managed throughout the change process or project lifecycle.							

Information or Data Item	Requested YES-NO	Date Requested	Reference or ID Number	Date Provided	Priority	Impact on M&A (e.g. Hi, Med, Low)	Summary or Comments
Request any studies of how these planned changes will affect the overall risk profile of the site as well as the business as a whole.							
Request information demonstrating all necessary approvals, where applicable, for these changes have been obtained or where approvals are outstanding, the reasons why and an estimated time to obtain the approval.							
Where an approval is outstanding, request information of the impact it might have on the project or change.							

M&A P.S. CHECKLIST – PLANNING THE SITE VISITS

Issue or Activity	Date	Summary of Findings	Key Issues to be investigated
Once the location(s) of the site(s) are known, "Google Earth"® each site to obtain an overview of the site and its immediate environs.			
Note issues that should be investigated during the site visit especially any variances between the data supplied in the Data Room and that obtained from other sites.			
Conduct searches for any regulatory infractions, notices of violations, etc. on government websites (e.g. US-OSHA, EPA, Dept of Justice as well as state sites, UK-Health and Safety Executive or Environment Agency, European Union, etc.)			
Conduct searches of local newspapers or regional news services for reports of process safety or major hazard related events.			

Issue or Activity	Date	Summary of Findings	Key Issues to be investigated
Identify where records (e.g. P&ID's, maintenance & inspection records) are stored. Where possible review prior to site visit as part of preparations.			
Agree whether photos of the sites and equipment can be taken as well as whether interviews can be recorded. If yes, organize and take along cameras, voice recorders, etc.			
Identify all safety requirements for each site. Where practicable take along and arrive with all necessary PPE and arrange for any required safety orientations.			
Once the location(s) of the site(s) are known "Google Earth"® each site to obtain an overview of the site and its immediate environs.			
Note issues that should be investigated during the site visit especially any variances between the data supplied in the Data Room and that obtained from other sites.			

M&A P.S. – ISSUES TO BE INVESTIGATED DURING THE SITE VISITS

Issue or Activity	Date of Visit	Addressed Yes-No	Summary of Findings	Priority (e.g. a regulatory need)	Impact on M&A (e.g. Hi, Med, Low)	Recommended Follow-Up & Responsibility
General History of the Site(s)						
Age of facility(ies)						
History of ownership						
Materials used or manufactured previously at the site(s) such as toxic, carcinogenic, mutagenic, pyrophoric, oxidizing, etc.						
Previous land usage (e.g. solid or liquid waste disposal site)						
Previous issues or disputes with neighbors (e.g. noise, noxious fumes or vapors, etc.)						
Type of regular community interaction with:						
Local politicians						
NGO's if any						
Directly with 'neighbors'						

Issue or Activity	Date of Visit	Addressed Yes-No	Summary of Findings	Priority (e.g. a regulatory need)	Impact on M&A (e.g. Hi, Med, Low)	Recommended Follow-Up & Responsibility
Are all current regulatory permits in place and dates these expire?						
General Location						
Access to site by road, rail, sea, etc.						
Is the site subject to natural events (e.g. flooding, hurricanes, earthquake, etc.)?						
Type of industry in the surrounding area(s).						
Proximity to people or environmentally sensitive areas;						
Adjacent land usage (e.g. farming, tourism, schools, hospitals, parks, etc.)						
Prevailing wind patterns,						
Drainage of the facility land to surrounding areas.						
Site Documentation						
Review P&ID's, process safety information, etc. kept on site.						

Issue or Activity	Date of Visit	Addressed Yes-No	Summary of Findings	Priority (e.g. a regulatory need)	Impact on M&A (e.g. Hi, Med, Low)	Recommended Follow-Up & Responsibility
Is this information current and readily accessible to those who need it?						
Examine a random set of inspection, maintenance and test records and assess if equipment is being properly maintained						
Examine operating and maintenance procedures.						
Are they current and readily available?						
Review a random sample of MOC's for completeness. Compare against actual physical equipment/plant.						
Buildings and Offices on the Site/Facility						
Their proximity to process area(s),						
Has a facility siting analysis been conducted on all occupied						

Issue or Activity	Date of Visit	Addressed Yes-No	Summary of Findings	Priority (e.g. a regulatory need)	Impact on M&A (e.g. Hi, Med, Low)	Recommended Follow-Up & Responsibility
buildings?						
Verify whether any occupied buildings (temporary or permanent) are located within the predicted damage zones.						
What site or local operational controls are in place to assure temporary structures are not placed in damage zones?						
Verify the status of all recommended actions or changes as contained in the facility siting studies.						
Building construction (age & current condition of all occupied buildings),						
Extent of blast resistance (where necessary) or construction to limit damage						

Issue or Activity	**Date of Visit**	**Addressed Yes-No**	**Summary of Findings**	**Priority (e.g. a regulatory need)**	**Impact on M&A (e.g. Hi, Med, Low)**	**Recommended Follow-Up & Responsibility**
Whether any hazardous process(es) are located in enclosed buildings,						
Are there any operations that generate dust and are there programs in place to control such hazards,						
Egress from and access to facilities in the event of an emergency,						
Condition and location of control rooms,						
Condition and location of warehouses,						
Control and Instrumentation systems						
Review any Safety Integrity Level (SIL) analyses of the process and safety control systems.						
Extent of instrumentation:						
Central monitoring-local control vs. central monitoring and						

Issue or Activity	Date of Visit	Addressed Yes-No	Summary of Findings	Priority (e.g. a regulatory need)	Impact on M&A (e.g. Hi, Med, Low)	Recommended Follow-Up & Responsibility
control						
Degree of automation, computerization and redundancy in system(s)						
Number, types and age of control systems in place						
Fire and gas detection systems:						
Alarm only,						
Automatic activation						
General condition of physical plant, equipment and machinery						
Separation distances between process areas, trains, tanks, fired heaters, etc.						
Identify the approximate (total) numbers of:						
Pressure vessels,						
Storage tanks,						
Pressure relief valves						
General state of process piping,						
Any underground process piping,						

Issue or Activity	Date of Visit	Addressed Yes-No	Summary of Findings	Priority (e.g. a regulatory need)	Impact on M&A (e.g. Hi, Med, Low)	Recommended Follow-Up & Responsibility
General state of insulated piping, vessels, etc. (visible water dripping, seepage)						
General appearance of all process vessels, tanks, etc.						
General appearance and/or noticeable noises, temperatures, etc. of rotating equipment						
General appearance or condition of process instrumentation connections to process piping, vessels, etc.						
General appearance or condition of relief valves, blow-down headers, flares, etc.						
General appearance of electrical terminations in hazardous areas and seals where cables penetrate control room walls.						

Issue or Activity	**Date of Visit**	**Addressed Yes-No**	**Summary of Findings**	**Priority (e.g. a regulatory need)**	**Impact on M&A (e.g. Hi, Med, Low)**	**Recommended Follow-Up & Responsibility**
General overall housekeeping of the facility						
Storage of flammable and hazardous materials						
Separation of storage tanks from process areas, boundary limits, fired heaters, etc.						
Types of storage tanks used (floating roof, cone, etc.)						
Condition of any tank farms (dikes, drainage via safe drain system, fire protection, access/egress, etc.)						
Adequate to contain a worst-case release?						
Any underground storage or underground storage tanks.						
Room for the expansion of any tank farms in the future						
Drum storage						

Issue or Activity	Date of Visit	Addressed Yes-No	Summary of Findings	Priority (e.g. a regulatory need)	Impact on M&A (e.g. Hi, Med, Low)	Recommended Follow-Up & Responsibility
Location of drum storage areas in relation to other plant, equipment, etc.						
How are drums off-loaded, moved around the facility, etc.						
Drainage via safe drain system						
Warehousing of packaged hazardous materials						
If organic peroxides are stored, are they being stored in air-conditioned warehouses?						
Are any warehouses attached to process buildings? Do they contain flammable or combustible fluids and if so is there warehouse designed to safely capture and drain any spills?						
Fire protection system						
Type and general condition of the						

Issue or Activity	Date of Visit	Addressed Yes-No	Summary of Findings	Priority (e.g. a regulatory need)	Impact on M&A (e.g. Hi, Med, Low)	Recommended Follow-Up & Responsibility
site fire alarm system						
Fire water sources						
Any fire water tanks and their general state						
General state of any deluge/foam systems						
General state of fire water pumps						
Provisions for retention of fire water						
General philosophy of the fire, smoke and gas detection system(s) (area, release source, etc.)						
Are the actual locations of detectors in accord with this philosophy?						
Ascertain the general condition of the fire water monitors, sprinklers, foam systems, etc.						
Confirm employee alarm systems are in place, functional and periodically tested.						

Issue or Activity	Date of Visit	Addressed Yes-No	Summary of Findings	Priority (e.g. a regulatory need)	Impact on M&A (e.g. Hi, Med, Low)	Recommended Follow-Up & Responsibility
Site or facility fire brigade and state of any required equipment.						
Do on-site emergency response teams receive a minimum of 40hrs training and three days of actual field experience?						
Utilities and services						
General state of utility systems and ancillary equipment such as cooling towers, drains, plant air, etc.						
Adequacy and source of power supplies for current and future needs and reliability of such sources.						
General condition of electrical supply equipment						
Emergency power available						
Site wide communication						

Issue or Activity	**Date of Visit**	**Addressed Yes-No**	**Summary of Findings**	**Priority (e.g. a regulatory need)**	**Impact on M&A (e.g. Hi, Med, Low)**	**Recommended Follow-Up & Responsibility**
systems						
Internal phones, radios, etc.						
Liquid waste disposal system and general state.						
Adequacy and condition of any steam systems						
Utilities supplied underground and ever inspected?						
Transfer or loading-unloading plant and equipment						
Pipelines						
Pipelines that enter or exit the facility as well as pipelines located near to the property boundaries.						
What are the regulatory requirements or jurisdiction (state, federal, etc.) for the operation of any associated pipelines?						
Past inspections by the appropriate regulatory agency(ies) and status of their findings.						

Issue or Activity	Date of Visit	Addressed Yes-No	Summary of Findings	Priority (e.g. a regulatory need)	Impact on M&A (e.g. Hi, Med, Low)	Recommended Follow-Up & Responsibility
Age of the pipeline(s), history of ownership, previous service, maximum allowable operating pressure, etc.						
Location of the control center and its ability to detect malfunctions, leaks, releases, etc. and effectively control such events.						
Review the right of way of the pipeline for areas of high consequence.						
Records of past issues concerning the right-of-way with landowners, local authorities, etc.						
History of any leaks or other releases.						
Status of most recent inspections and testing, cathodic protection systems, corrosion inhibition/injection, etc.						

Issue or Activity	Date of Visit	Addressed Yes-No	Summary of Findings	Priority (e.g. a regulatory need)	Impact on M&A (e.g. Hi, Med, Low)	Recommended Follow-Up & Responsibility
Docks						
If the facility receives or transports material via marine docks or jetties what is the extent of Safety and Environmental controls the facility exerts over those operations?						
What is the regulatory status of the dock or jetty (e.g if under US Coast Guard authority – is there a current "Letter of Adequacy")?						
What are the materials transported by dock or jetty?						
Types, size and nature of vessels that use the docks,						
Are the materials loaded in bulk, by drum, tank-tainer, etc.?						
Where materials are loaded in bulk, does a surveyor inspect the vessel and all required certificates prior to loading?						

Issue or Activity	Date of Visit	Addressed Yes-No	Summary of Findings	Priority (e.g. a regulatory need)	Impact on M&A (e.g. Hi, Med, Low)	Recommended Follow-Up & Responsibility
Quantities shipped and the frequency of loading/unloading.						
Potential consequences of releases to the marine environment.						
What emergency or contingency plans are in place and how often are these actually exercised.						
Loading Racks (Road and Rail Tanker)						
Location of loading racks and their proximity to process trains, tank farms, and off-site facilities.						
What are the materials loaded, quantities, frequency of loading, etc.						
Access and egress to from loading racks.						
Access and egress in the event of an emergency.						

Issue or Activity	**Date of Visit**	**Addressed Yes-No**	**Summary of Findings**	**Priority (e.g. a regulatory need)**	**Impact on M&A (e.g. Hi, Med, Low)**	**Recommended Follow-Up & Responsibility**
Nature and condition of fire detection and fire fighting systems.						
Nature of the instrumentation and control systems.						
Condition of the loading arms.						
Where hoses are used, program to assure their integrity.						

M&A P.S. CHECKLIST – PROCESS SAFETY ISSUES TO BE CONSIDERED

Issue or Activity	Date	Summary of Findings	Priority (e.g. a regulatory need)	Impact on M&A (e.g. Hi, Med, Low)	Recommended Follow-On Actions	Responsible & Time to Complete
If processes at the site fall within the scope of any specific requirements for process safety reviews, risk management plans, formal safety assessments and/or Safety Cases to what extent is the site in compliance?						
For sites, or portions of sites not covered by process safety or major accident hazard regulations what alternate types of Process Safety related reviews, hazard analyses, etc. are performed?						

Issue or Activity	Date	Summary of Findings	Priority (e.g. a regulatory need)	Impact on M&A (e.g. Hi, Med, Low)	Recommended Follow-On Actions	Responsible & Time to Complete
If the site has been the subject of any process safety or major accident hazard reviews or audits what were the findings?						
What remains open?						
To what extent is documentation of closed items sufficient?						
What practices are in place to ensure equipment, including utilities, is properly designed for the intended service, inspected and adequately maintained?						
How effectively are these practices applied?						

Issue or Activity	Date	Summary of Findings	Priority (e.g. a regulatory need)	Impact on M&A (e.g. Hi, Med, Low)	Recommended Follow-On Actions	Responsible & Time to Complete
What are the qualifications of inspectors and do they meet any regulatory or industry standards?						
During construction of modifications what processes were used to ensure that equipment, piping, etc. met the specifications in the design for safe operation?						
What engineering standards were applied in design and subsequent maintenance of the operations?						
Are the standards used in conformance with internationally recognized practice, standards or regulatory requirements?						

Issue or Activity	Date	Summary of Findings	Priority (e.g. a regulatory need)	Impact on M&A (e.g. Hi, Med, Low)	Recommended Follow-On Actions	Responsible & Time to Complete
How well does available information support that these standards are being complied with?						
How have pressure relief systems, including associated lines, flares, etc. been designed and evaluated?						
What open issues exist?						
To what extent are systems in compliance with applicable pressure vessel codes for design, inspection and maintenance?						
Note any relief vents to atmosphere and assure dispersion studies have been undertaken to verify they are vented to a safe location.						

Issue or Activity	Date	Summary of Findings	Priority (e.g. a regulatory need)	Impact on M&A (e.g. Hi, Med, Low)	Recommended Follow-On Actions	Responsible & Time to Complete
What is the management's system for assessing required process safety issues and whether these adequately address all potentially foreseeable PS events?						
To what extent is the data available regarding process chemistry sufficient to identify all relevant significant hazard scenarios, such as potential runaway reactions (where appropriate)? How has it been assessed to ensure its robustness?						

Issue or Activity	**Date**	**Summary of Findings**	**Priority (e.g. a regulatory need)**	**Impact on M&A (e.g. Hi, Med, Low)**	**Recommended Follow-On Actions**	**Responsible & Time to Complete**
What is the site history of incidents which are of a type targeted for prevention through process safety reviews and management of change processes?						
What exercises and tests have been conducted to ensure the adequacy of emergency systems for mitigating releases or controlling fires? What recommendations have arisen from such exercises and have these been effectively acted on?						
To what extent is documentation on file to verify that Process Safety programs are in place and effective?						

M&A P.S. CHECKLIST – ASSESSING MAJOR HAZARD RISKS

Issue or Activity	Date	Summary of Findings	Recommended Follow-On Actions	Responsible	Estimated Time to Complete
Materials Used and/or Products Handled					
What raw materials are used?					
How are raw materials supplied (pipeline, road or rail, etc.)					
Storage quantities and practices.					
What combinations of substances and quantities, location and proximity to people or environmental receptors exist which pose a credible risk of a major accident?					
What intermediates are produced?					
Any issues with intermediates?					
What products are produced?					

Issue or Activity	Date	Summary of Findings	Recommended Follow-On Actions	Responsible	Estimated Time to Complete
How are they supplied to customers?					
Methods of shipment					
Quantities Shipped					
What are the results of any qualitative and quantitative risk assessments to employees, contractors, the public and environment?					
What are the scenarios on which these assessments are based?					
To what extent was the process for evaluating major accident risks validated?					
What programs and objectives are in place to achieve continuous reductions in risks posed by the operations?					

Issue or Activity	Date	Summary of Findings	Recommended Follow-On Actions	Responsible	Estimated Time to Complete
How has the major accident risk assessment information been shared with appropriate government authorities and the public?					
Where such information must be provided by regulation, how was this accomplished and how is that information kept current?					
How do the risks compare with applicable standards?					
Have there been any process safety or major hazard related incidents or accidents at any of the site(s) in the past X years?					

Issue or Activity	Date	Summary of Findings	Recommended Follow-On Actions	Responsible	Estimated Time to Complete
What is the status of any findings or recommendations arising from these incidents?					

M&A P.S. CHECKLIST - PROCESS SAFETY MANAGEMENT & CULTURE

Issue or Activity	Date	Summary of Findings	Recommended Follow-On Actions	Responsible	Estimated Time to Complete
Does the Process Safety record of the site or facility Operator pose a significant reputational risk?					
What is the history of Process Safety related claims, compliance, etc. of the site Operator?					
Are there any court orders, waivers, notices of violations, administrative orders in place that pertain to the site or operations?					
What is the nature and structure of the business' process safety management system?					

Issue or Activity	Date	Summary of Findings	Recommended Follow-On Actions	Responsible	Estimated Time to Complete
If a periodic HSE report is produced for management assurance or public dissemination, does it address Process Safety and what was contained in recent reports?					
Assess process safety accruals/reserves:					
To what extent is the Management System credible in assessing future process safety liability?					
How well does any existing accrual or reserve match probable estimable liability(ies)?					
To what extent is the calculation of reserves auditable? What was found in the most recent audit?					

Issue or Activity	Date	Summary of Findings	Recommended Follow-On Actions	Responsible	Estimated Time to Complete
What concerns, other than soil and groundwater at owned sites are covered by reserves?					
What written instructions or policies are in place regarding process safety matters?					
How effectively are they enforced?					
What gaps or weaknesses impair their effectiveness?					
For those facilities required to have formal management safety systems, such as a 'Safety Case', to what extent are these current and actively used as management tools?					

Issue or Activity	Date	Summary of Findings	Recommended Follow-On Actions	Responsible	Estimated Time to Complete
If significant gaps exist between the formal program and current or expected conditions, what degree of effort is needed to address any variances?					
Where regulatory review or approval of some or all of the Management System is required, what is the status of any such review(s)?					
What were the findings of any internal or third party compliance assessments? What follow-up actions were taken and plans put in place as a result?					
What is the facility/company ISO9000, 14001 and 18001 status? Especially in regard to how					

Issue or Activity	Date	Summary of Findings	Recommended Follow-On Actions	Responsible	Estimated Time to Complete
Process Safety matters are integrated into such programs?					
Overall, what is your subjective judgment of the Process Safety management system?					
How well prepared is the organization to know what it needs to do and to ensure the Process Safety needs are being met?					
Are they making appropriate commitments and carrying through those commitments? Is the system sustainable?					
How much work and time is needed to get wherever they are to where they need to be?					

Issue or Activity	Date	Summary of Findings	Recommended Follow-On Actions	Responsible	Estimated Time to Complete
What kind of Process Safety culture exists in the organization?					
Evaluate the level of supervisory-management commitment to Process Safety.					
Any executive, senior and line management training on process safety issues provided or required?					
What is the communication style and practice of the overall organization as well as at the various sites and facilities?					
Is the communication style one-way and vertical creating potential silos of information or more of a networked style?					
Do staff (e.g. operators, maintenance					

Issue or Activity	Date	Summary of Findings	Recommended Follow-On Actions	Responsible	Estimated Time to Complete
craft) feel they are empowered to take corrective actions when a P.S. condition develops?					
Does staff feel they are encouraged and have the support of their direct line supervisor and managers to report near misses, safety interlocks being bypassed, etc.?					
Does staff feel they are kept well informed of Process Safety issues that affect their areas or may impact them?					
Does staff feel the organization is proactive or reactive in addressing process safety related issues?					
Is the culture of the organization an impediment to					

Issue or Activity	Date	Summary of Findings	Recommended Follow-On Actions	Responsible	Estimated Time to Complete
achieving the desired Process Safety performance?					
What kind of changes will be needed and what level of effort is required to accomplish this?					

M&A P.S. CHECKLIST – PROCESS SAFETY STAFFING ISSUES

Issue or Activity	Date	Summary of Findings	Recommended Follow-On Actions	Responsible	Estimated Time to Complete
What is the organization and responsibilities for Process Safety?					
How is Process Safety support currently provided?					
Wholly internal or combination of internal/external?					
Evaluate:					
Level of expertise provided					
Potential gaps					
Continuity in the staff					
Assess any traditional HSE interactions with Process Safety issues.					
What is the relationship with outside authorities?					

Issue or Activity	Date	Summary of Findings	Recommended Follow-On Actions	Responsible	Estimated Time to Complete
How will staffing needs and availability of staff be impacted by the transaction?					
What corporate (or other off-site) process safety personnel and systems are used?					
What transition services will be needed and how will they be provided?					

M&A P.S. CHECKLIST – HAZARD IDENTIFICATION ISSUES TO EVALUATE

Issue or Activity	Date	Summary of Findings	Recommended Follow-On Actions	Responsible	Estimated Time to Complete
Review the facility's approach of systematically identifying, evaluating and controlling processes involving highly hazardous chemicals.					
Does a plan exist for conducting PHA's such that all processing areas have had at least a baseline PHA performed?					
Are PHA's conducted using teams of personnel from operations and technical and is led by an individual knowledgeable in the methodology?					
Is a system is in place to address all findings of PHA's?					

Issue or Activity	**Date**	**Summary of Findings**	**Recommended Follow-On Actions**	**Responsible**	**Estimated Time to Complete**
Are PHA's updated every five years?					
Is there a hazard and risk register in place?					
How is this used?					
Is senior and executive management regularly briefed on the status of hazards and risks?					

M&A P.S. CHECKLIST – MANAGEMENT OF CHANGE ISSUES TO INVESTIGATE

Issue or Activity	Date	Summary of Findings	Recommended Follow-On Actions	Responsible	Estimated Time to Complete
What is the site's process for management of change? How effective is the MoC process in practice?					
Review the facility practices for evaluating and approving all modifications to equipment, raw materials and processing conditions other than replacement in kind.					
Assure procedures address the following:					
The technical basis for the change is evaluated,					
Impact on safety and health is evaluated,					

Issue or Activity	Date	Summary of Findings	Recommended Follow-On Actions	Responsible	Estimated Time to Complete
Changes to operating procedures are made,					
How temporary and emergency changes are addressed,					
How the over-ride or bypassing of safety interlocks is managed and controlled,					
Necessary time period for change is addressed,					
Authorizations are properly obtained,					
All documents associated with the change (e.g. P&IDs, operating and maintenance procedures), inspection, testing procedures, etc. are revised in accordance with the change.					

Issue or Activity	Date	Summary of Findings	Recommended Follow-On Actions	Responsible	Estimated Time to Complete
Examine the written procedures for managing technological and organizational changes.					
Are they in accord with current industry practice?					
Are they effectively followed when such changes are made?					
Is there a backlog of open MoC's?					
Is the backlog tracked and periodically provided to management?					

M&A P.S. CHECKLIST – MECHANICAL INTEGRITY ISSUES TO INVESTIGATE

Issue or Activity	Date	Summary of Findings	Recommended Follow-On Actions	Responsible	Estimated Time to Complete
Review historical spending for capital and maintenance.					
Are there discernible trends?					
Is there evidence of an obvious under-spend?					
What is the general form of the inspection and maintenance management program – e.g. preventative, condition based, reactive, etc.?					
Does the facility use a risk based inspection program?					
Does the RBI program follow recognized industry standards or practice (i.e is it in accord with RAGAGEP)?					

Issue or Activity	Date	Summary of Findings	Recommended Follow-On Actions	Responsible	Estimated Time to Complete
Review facility practices and procedures to identify critical equipment.					
Determine if all critical equipment is being inspected and maintained per requirements.					
Review the facility's practices and systems used to track and prioritize work requests (e.g. is a Computer Maintenance Management Systems-CMMS).					
Review procedures to track work orders. coordinate maintenance work activities, manage schedule changes.					

Issue or Activity	Date	Summary of Findings	Recommended Follow-On Actions	Responsible	Estimated Time to Complete
Review facility procedures to track all equipment or systems requiring preventive or predicative maintenance.					
Review facility practices for approving delays in scheduled maintenance or critical function testing on instrumentation.					
Is a log of all overdue tests/inspections maintained?					
Are there any outstanding or overdue tests or inspections of Safety Critical equipment?					
Is management periodically apprised of overdue tests and inspections?					

Issue or Activity	Date	Summary of Findings	Recommended Follow-On Actions	Responsible	Estimated Time to Complete
Review the administration and control of spares.					
Review facility procedures or practices to plan and execute maintenance turnarounds.					
During maintenance, what is the process for ensuring replacement parts are suitable and actually meet specifications?					
How are maintenance and construction contractors vetted to ensure competency?					

M&A PS CHECKLIST – PROCESS SAFETY ISSUES TO EXAMINE

Issue or Activity	Date	Summary of Findings	Recommended Follow-On Actions	Responsible	Estimated Time to Complete
Review the facility practices for keeping the following information current; PFD's; P&ID's; process chemistry, physical reactivity and corrosivity, toxicity and safe upper/lower operating conditions.					
Evaluate the following:					
Whether the required written process safety information is provided to all PHA teams prior to their start,					

Issue or Activity	Date	Summary of Findings	Recommended Follow-On Actions	Responsible	Estimated Time to Complete
Whether MSDS information is available to operators, maintenance staff, contractors, etc. who work with hazardous materials,					
Whether documentation exists that verifies equipment is fit for service or complies with applicable recognized and generally accepted good engineering practices or good industry practice.					

M&A PS CHECKLIST – PROCESS SAFETY PROCEDURES TO EXAMINE

Issue or Activity	Date	Summary of Findings	Recommended Follow-On Actions	Responsible	Estimated Time to Complete
Review the site Process Safety procedures or manual or where relevant the site Safety Case and confirm current and readily available to all relevant staff.					
Review any Operating and Maintenance manuals to confirm they are current and readily available to all relevant staff, contractors, etc.					
Review a selection of operating logs for:					
Process safety concerns or hazardous conditions,					

Issue or Activity	Date	Summary of Findings	Recommended Follow-On Actions	Responsible	Estimated Time to Complete
Notations where the operations may have been near to or in excess of critical operating parameters.					
Notations where interlocks are bypassed or disabled.					
Review how the facility assures operators are competent and where required certified to carry-out their assigned tasks and duties.					
How is training scheduled, tracked and monitored by management? Is a computerized system used?					
Review the emergency shutdown procedures.					
Are the responsibilities and authority for shutting down a hazardous situation clearly delineated?					

Issue or Activity	Date	Summary of Findings	Recommended Follow-On Actions	Responsible	Estimated Time to Complete
Are the procedures current and readily accessible to all who need them?					
Do the emergency shutdown procedures align with the major accident scenarios developed in the Process Hazard Analyses - Formal Safety Assessments?					
Are the emergency shutdown procedures tested regularly as part of scheduled emergency drills?					

M&A P.S. CHECKLIST – P.S. AUDIT ISSUES TO CONSIDER

Issue or Activity	Date	Summary of Findings	Recommended Follow-On Actions	Responsible	Estimated Time to Complete
Are periodic internal audits conducted?					
Review past audits of the facility's Process or major hazards safety program.					
Have any external audits been conducted by regulatory groups, insurance carriers, joint venture partners, external contractors?					
Do senior managers receive the findings and recommendations of audits?					
How are recommendations from such audits, tracked, resolved and closed-out?					
How is responsibility for closing out recommendations assigned?					

Issue or Activity	Date	Summary of Findings	Recommended Follow-On Actions	Responsible	Estimated Time to Complete
Do those assigned such responsibility have the authority and resources to address effectively the finding(s)?					
Verify that a selective sample of audit recommendations have been resolved and properly closed-out.					

Appendix B – An Exemplar Integration Plan & Budget

GUIDANCE FOR USING THE PLAN AND BUDGET SPREADSHEETS

Background & Instructions: Exemplar Integration Plan and Budget
Background Information Regarding the Development of the Spreadsheets
1) The tabs below will open up two different spreadsheets that were developed to assist a reader with preparing their own integration plan and a budget or estimate of the resources necessary to implement that plan.
2) Before opening the spreadsheets, a few words on their background and development might help the reader put them into context with respect to their own situation.
3) The spreadsheets are a build-out or enhancement of an actual integration plan developed for an acquisition and merger ot two relatively like companies. When the companies combined their total market capitalization was put at US$38 Billion.
4) Both of the companies were multi-nationals and had operations spread-out around the world.
4) However, after the deal closed and detailed reviews of the manner by which the two companies had approached HSE was started it was found their approaches were fundamentally different. One was very rules based, the other risk based.
5) Further differences in the manner by which HSE in general, and process safety issues in particular began emerging as the post closing reviews continued.
6) By the time the reviews were complete, it was evident there were considerable differences or variations between the two approaches to process safety and the various program elements.
7) One boundary condition set for the integration though was full integration and a common program. However, it was recognized both companies' programs had good elements and the new program was to encapsulate all good elements.
8) There were regulatory pressures in the background as well.
9) It should also be noted that the executives from company with the overall stronger program of the two, ended up in key operational and HSE positions. One of those key executives had a personal passion for Process Safety matters.

Background & Instructions: Exemplar Integration Plan and Budget
Background Information Regarding the Development of the Spreadsheets (continued)
10) There was solid executive support for a well planned and executed integration process. However integration was to be completed within three years of the acquisition (i.e. the process was 'front end loaded')
11) As your work through the two spreadsheets then, bear in mind the concept of 'proportionality'. A mega-merger might well entail all that is contained in the spreadsheets both in what and how to achieve it.
12) The acquisition of a single-site or operation, however, is unlikely to require the effort(s) and investments as outlined in the spreadsheets. That said it is generally easier to delete rows and columns than add such material.
Exemplar Integration Plan
1) The plan lays-out a series of activities for identifying, mobilizing as well as assigning possible tasks for a series of teams, details of which are laid-out in Chapters 5 and 6.
2) The spreadsheet was developed to cover a calendar year (52 weeks) for the integration process. While it is divided into 13 months this is only because it is not possible to determine in any-one quarter when a 5-week month occurs.
3) While there are various project management software packages available for developing such a plan, Excel is more commonly available and that is why Excel was chosen over using a special PM package.
4) If it is desired to develop an overall Gantt Chart, the reader has the option of filling the cell(s) with a color or pattern or turning on 'draw' and drawing in various lines or arrows.
5) The spreadsheet was not developed as a prescriptive set of activities nor prescribing a certain formation of teams required to deliver a successful integration process. It is provided as a foundation on which to build your own individual integration plan.
6) In simple terms, cut, paste, add, revise, delete to suit your particular needs and constraints.
Exemplar Integration Budget
1) The Exemplar budget spreadsheet was developed to accompany the Integration Plan. It will calculate the human resources investment both in terms of hours and potential costs.
2) You will have to do a little interpolation between the Plan and the Budget as the plan is more activity focused while the budget is more personnel or team based focused.

Background & Instructions: Exemplar Integration Plan and Budget
Exemplar Integration Budget
3) The spreadsheet consists of three general parts. The first part provides the ability to estimate the investment in terms of hours.
4) The second part will take those hours and calculate a cost. The third part provides the ability to estimate associated expenses and sum those into an overall total.
5) Before loading any data into the weekly columns for a particular individual, you should first scan over to column "BD" and load in estimates of the cost rate for a particular position or individual.
6) Some guess-estimates are already loaded into column "BD", these should be replaced by rates that are applicable to your organization or integration process.
7) Once you have loaded in applicable cost rates, you can then start filling in the weekly columns for a particular position or individual.
8) The spreadsheet will then start calculating a weekly total of hours invested as well as an annual total hours invested for that particular individual or position.
9) In the second part, the costs, the spreadsheet will automatically multiply those hours by the appropriate cost figure and total those amounts again on a weekly as well as annual basis.
10) Grand totals are also automatically calculated.
11) The Total Hours are given in two columns - column 'C' and 'BE'. This was done only to provide the convenience of being able to keep an eye on the total whether working at the 'front' or 'back' of the spreadsheet.
12) When deleting or inserting a row make sure you delete or insert that same row in all three parts of the spreadsheet, to keep the calculations accurate.
13) Further when inserting a row in the second part of the spreadsheet you will need to copy down all of the calculations in the row directly above into the new line you inserted. If you don't the monetary costs will not be calculated in that new row.
14) The third part that calculates associated expenses is not as integrated with the first two parts. It will sum all the expenses in an individual week as well as annual row total.
15) The spreadsheet will also total the human resource related investments with the associated fees to provide you with an overall grand total.

AN EXEMPLAR INTEGRATION PLAN

TASKS	Months and Weeks 'Beginning'											
	Month 1				Month ...				Month 12			
	1	2	3	4	...	...	...	...	50	51	52	53
Establish Selection Process for identifying & appointing Implementation Teams												
Appoint Process Safety Implementation Sponsor(s) from Executive Team												
Appoint & Charter Guiding Team *(i.e. A **working** Steering Committee)*												
Appoint a Project Director (because of the size, suggest a 'director' over a project manager)												
Appoint a Project Secretariat *(see below for further explanation)*												
Identify Implementation Teams (e.g. 'intolerable risks', physical plant & equipment, management systems, etc.)												
Develop Selection Criteria for potential members of the Implementation Teams												

TASKS	Months and Weeks 'Beginning'											
	Month 1				Month …				Month 12			
	1	2	3	4	…	…	…	…	50	51	52	53
Identify and appoint Implementation Team Leads (to be done jointly by P.S. Impl Lead & Guiding Team)												
Develop and agree a selection process for implementation team members												
Secretariat to contact potential members and their management to determine interest & availability												
P.S. Impl Lead to develop draft team charters and obtain Guiding Team approval												
Collect and screen potential candidates against Criteria												
Interview and appoint Team Members												
Guiding Team and All Implementation Teams hold a 1/2 to 1 day 'kick-off' retreat to assure alignment												
Establish a Communications Process with various Stakeholders (Exec, Line Managers, Staff, Contractors, etc)												
Agree Executive level interface												

TASKS	Months and Weeks 'Beginning'											
	Month 1				Month ...				Month 12			
	1	2	3	4	...	...	...	...	50	51	52	53
Agree Executive interface authority (e.g. communications only or comms plus executive power)												
Consider establishing a 'stakeholder liaison group' (e.g. Industry Stakeholders, Unions, Local Community etc.)												
The P.S. Impl Lead & Team Leads to provide 'periodic' reports to Stakeholder Liaison												
Prepare internal press release on the Team charters and Team members *(release externally?)*												
Consider arranging an internal interview of the P.S. Impl Lead & Team Leads for internal release												
Appoint 'Single Points of Contacts' (SPOC's) for Key Groups, Functions, Business Units, Plants or Facilities, etc.												
These would include the individual plants or sites, maintenance, Operations Excellence, IT, etc.												

TASKS	Months and Weeks 'Beginning'											
	Month 1				Month ...				Month 12			
	1	2	3	4	...	...	...	...	50	51	52	53
Identify other groups the Implementation Teams may need support from or impact - and appoint SPOC's												
Consider establishing a Secretariat for the Implementation Process												
Evaluate the total administrative burden that is likely to arise to support the Guiding Team, the P.S. Impl Lead as well as the various Implementation Teams. The usual approach is to require the various groups generally 'to beg, borrow and/or steal' such resources from wherever they can. In the end this generally leads to frustration from all parties and can turn-out to be the limiting factor of the implementation activities making progress as a whole. Consider appointing a Secretariat to the implementation process to support these needs. In establishing such a Secretariat a critical factor will be their knowing where and how to find and access various materials, documents and information.												
Appoint a 'Secretariat' (i.e. Project Administrator or Co-ord) possibly independent from any existing P.S. group												
Secretariat to identify and secure potential 'staffing needs' if any												
Identify and secure necessary needs (e.g. computers, file space, etc.)												
Implementation Teams Develop their individual Working Processes												

TASKS	Months and Weeks 'Beginning'											
	Month 1				Month …				Month 12			
	1	2	3	4	…	…	…	…	50	51	52	53
Detailing how the Implementation Teams are to go about their work in order to discharge their duties has the potential of providing a confusing message where on the one hand you are empowering them, then turning around and specifying how they are to go about undertaking what is required of them. That said it is often useful to help kick-start a change process as involved in an integration by providing a few concrete examples and immediate steps in addition to providing a broad picture and general direction for the Authority as a whole. The following set of activities has been developed more to assist with developing schedules and planning than to suggest the actual approach the teams may wish to adopt and implement.												
Develop and agree needs, approach(es) and required resources												
P.S. Impl Lead holds initial meeting and planning workshop to clarify objects and immediate deliverables												
Key issues agreed with the Teams												
General plan agreed with the Teams, and deliverables defined												
Schedule constraints (e.g. Holidays, Requirements per Admin Agreement, etc.) identified and agreed												
Teams examine current situation against desired state												
Gaps identified along with an approach (or possibly multiple approaches) to address gaps												

TASKS	Months and Weeks 'Beginning'											
	Month 1				Month …				Month 12			
	1	2	3	4	…	…	…	…	50	51	52	53
Estimate made of the time and resources required for each 'solution'												
Team develops initial 'solution plan' with first order estimates of required resources												
First order estimates collected and collated by P.S. Impl Lead												
P.S. Impl Lead presents compiled list of solutions & resource implications to Guiding Team												
Guiding Team, P.S. Impl Lead and Team Leads prioritize and agree what can be feasibly undertaken												
Additional work to refine resource requirements and or scope of solutions identified where necessary												
Timeline agreed for presentation of 'second order' resource estimates, where necessary												
Agree whether there is a need to supplement the Impl Teams to complete 2nd order estimates												
P.S. Impl Lead & Secretariat to identify and secure such skills, if required.												

TASKS	Months and Weeks 'Beginning'											
	Month 1				Month ...				Month 12			
	1	2	3	4	...	...	...	...	50	51	52	53
Teams undertake studies to refine both the approach and resource implications												
Second order estimates and work plans to P.S. Impl Lead												
P.S. Impl Lead collates all second order estimates												
Guiding Team, P.S. Impl Lead & Impl Team Leads agree second order estimates & work plans												
Second order estimates and work plans presented to Exec Sponsor												
Internal announcement of the final plans, resources required, etc.												
Consider release of work plans to all external stakeholders.												
Begin Implementation												
Note - Certain activities that preceded the M&A and can be dovetailed into the integration should continue. There is little doubt other 'projects or initiatives' will be agreed and phased in. In each case the appropriate implementation team should be required to estimate the potential resource needs through the end of week 53 and the status at end of week 53.												
Guiding Team Activities												

TASKS	Months and Weeks 'Beginning'											
	Month 1				Month ...				Month 12			
	1	2	3	4	...	...	...	...	50	51	52	53
'Nail Down' the Vision, Make it Come Alive & Tie it to the overall Implementation Plan for the business(es)												
Clear the decks of competing initiatives												
Identify all initiatives presently consuming required resources and those planned which could impact resources												
Rank criticality of all initiatives w.r.t. achieving the level of integration in process safety snr. Management desire												
Present initiative rankings vs. resource implications for snr management agreement or revision												
Put into place a process to address potential resistance from Leaders & Managers												
Develop a process to keep Line and key support management enrolled in the change												
Agree critical gaps between current services and 'desired state'												
Identify steps or actions required to address gaps												

TASKS	Months and Weeks 'Beginning'											
	Month 1				Month ...				Month 12			
	1	2	3	4	...	...	...	...	50	51	52	53
Estimate resources required to implement												
Monitoring of Implementation Progress, identification of possible changes or resource issues												
Identification of the need for possible changes to charters, scopes, activities												
Identification and corrective actions for changes in demands or need for resources												
Brief Executive Sponsor and obtain approval(s) if necessary												
The following teams and possible actions are put forward as a guide to help foster discussion and debate between the Process Safety Implementation lead, the Guiding Team and possibly the Executive Sponsor as to the formation of the actual teams and possible assignments for those teams. The material should be considered an 'Aunt Sally' or 'Straw Man' of possibilities. Further the actual process or means by which the individual teams will carryout their assignments should be left to those teams.												
'Intolerable' Risks Team												
Identify and agree critical or intolerable risks												
Develop scope of potential solutions & identify those which can be implemented within 90-100 days												
Estimate resources required to implement												

TASKS	Months and Weeks 'Beginning'											
	Month 1				Month ...				Month 12			
	1	2	3	4	...	...	...	...	50	51	52	53
Obtain approvals												
Implement immediate or near term controls												
Identify, evaluate and develop costed plans for long-term solutions												
Turn long term solutions over to the Physical Plant & Equipment team or others as appropriate												
Physical Plant, Equipment and/or Capital Works Team												
Review and continue the work of the 'intolerable' risks team												
Examine main processes and process related equipment												
Review Storage tanks, loading and offloading systems, etc.												
Examine critical control equipment												
Pressure relief systems,												
Fire & gas detection systems												
Instrumented control systems												
Fire containment and control systems												
Assess 'fitness for service' and compliance with industry and/or corporate engineering standards												

TASKS	Months and Weeks 'Beginning'											
	Month 1				Month ...				Month 12			
	1	2	3	4	...	...	...	...	50	51	52	53
Identify gaps and develop first order estimates of potential solutions												
Risk rank the various gaps, Identify solutions with long lead times												
Develop a phased budget and resourced corrective action plan												
Present plan to Senior Management Integration team or 'lead' and obtain approvals												
Revise plan if necessary												
Implement approved corrective action plan												
The Organizational Team												
Examine the organizational structures, seek out possible synergies.												
Analyze & identify activities critical to delivering or supporting process safety												
Determine interrelationship between critical activities and map to current structure(s)												
Seek out and identify potential 'synergies'												

TASKS	Months and Weeks 'Beginning'											
	Month 1				Month ...				Month 12			
	1	2	3	4	...	...	...	...	50	51	52	53
Based on desired state, assess required levels of authority, responsibility, communication flows, etc.												
Evaluate where and how activities and/or work steps can be effectively and efficiently grouped												
Map grouped activities to current structure												
Evaluate various options for merging supporting departments, groups or teams												
Liaise with Systems team to identify potential conflicts between systems and organizational structure												
Develop and incorporate plan(s) to minimize conflicts												
Liaise with Guiding team and Exec sponsor and develop plan to implement changes												
Lay ground-work with key stakeholders to implement change (i.e. Communicate, Communicate and then More)												
Implement changes												

TASKS	Months and Weeks 'Beginning'											
	Month 1				Month ...				Month 12			
	1	2	3	4	...	...	...	...	50	51	52	53
The Process and Systems team												
Is there a requirement to immediately integrate certain systems (e.g. accident/incident, monthly reports to management)												
Evaluate whether current in-situ systems can be modified or need to be replaced to meet requirements												
Develop first order estimates of costs and resources required to revise or replace systems												
Brief P.S. Implementation Lead, Guiding Team and Exec Sponsor on costs, obtain approvals or revise estimates												
Develop RFP's or P.O.'s to obtain necessary skills, resources or software packages required to implement changes												
Evaluate tenders and place orders, form project teams to assist implement the change(s)												

TASKS	Months and Weeks 'Beginning'											
	Month 1				Month ...				Month 12			
	1	2	3	4	...	...	...	...	50	51	52	53
Lay ground work for changes especially where changes impact day-to-day activities (e.g. changing reporting formats)												
Assess the need to revise the process safety document hierarchy of the newly acquired business(es)												
Address the need(s) to revise governing policy(s), strategies, integrity manuals, etc.												
Assess current state of second and tertiary procedures, guidelines, data sheets, etc. against desired state												
Assess the state of the current engineering standards against desired state.												
Identify gaps, develop resourced plan to transition these to the meet the expectation of the desired state												
Map existing processes and systems supporting delivery of the process safety program												
Identify gaps or requirements for changes between existing systems and those required for 'desired' state												

TASKS	Months and Weeks 'Beginning'											
	Month 1				Month …				Month 12			
	1	2	3	4	…	…	…	…	50	51	52	53
Determine need for new specific functions, reports, etc,												
Develop plans for any required changes (build on existing systems to the extent that it is effective and efficient)												
Liaise with Organizational team to identify potential conflicts between proposed changes and organizational structure.												
Liaise with Guiding team and Exec sponsor and develop plan to implement changes												
Lay ground-work with key stakeholders to implement change (i.e. Communicate, Communicate and then More)												
Implement changes												
The People Team												
Develop a RACI chart for P.S. in both the 'parent' organization and the newly acquired business(es)												
Identify variances between the two 'approaches' to process safety and the defined 'desired state'												

TASKS	Months and Weeks 'Beginning'											
	Month 1				Month ...				Month 12			
	1	2	3	4	...	...	...	...	50	51	52	53
Identify potential synergies in the merging of the process safety functions in liaison with the Organizational Team												
Present differences and 'synergies' to the Guiding Team and possibly the Executive Sponsor.												
Agree a synergies implementation plan with the Organizational Team, the Guiding Team and Executive Sponsor												
Implement 'synergies' plan.												
In liaison with the Systems & Organizational teams develop an on-boarding program for Key process safety stakeholders												
Program to reflect the vision and strategy for process safety in the newly combined businesses												
Program to identify at the '50,000' foot level changes that are likely to be required to align with the new vision and strategy												
Identify 'key stakeholders' (e.g. executive and senior managers, line												

TASKS	Months and Weeks 'Beginning'											
	Month 1				Month ...				Month 12			
	1	2	3	4	...	...	...	...	50	51	52	53
managers, key technical support, etc.)												
Roll-out of program												
Evaluate effectiveness of roll-out, identify areas of potential resistance												
Enlist aid of Guiding Team and Exec Sponsor to address areas of potential resistance												
With aid of key stakeholders develop a cascading 'on-boarding' process for all involved stakeholders to process safety												
Test roll-out on a representative sample of prospective stakeholders, identify areas for improvement & revise as necessary												
Schedule roll-out with SPOC's												
Roll-out of program												
Identify potential change champions among affected stakeholders as well as potential areas of resistance												
Assess and address culture towards P.S.												

TASKS	Months and Weeks 'Beginning'											
	Month 1				Month ...				Month 12			
	1	2	3	4	...	...	...	...	50	51	52	53
If not done previously, survey the culture towards process safety across the totality of the newly acquired businesses												
If a process safety cultural survey in the parent company has not been done, include one to assist with benchmarking												
Identify critical concerns or gaps in the current culture as well as areas of potential resistance												
Develop a program to address cultural attitudes and areas of potential resistance												
Remember program needs to address the 'hearts' more so than the 'minds'												
Engage identified 'change champions' in helping to roll the program out												
Engage individuals in positions of 'power' not necessarily organizational structure in the change program												
Roll program out through the use of the above 'change agents'												

TASKS	Months and Weeks 'Beginning'											
	Month 1				Month …				Month 12			
	1	2	3	4	…	…	…	…	50	51	52	53
Assess and possibly contrast the overall communication styles of the 'parent' and 'newly acquired business'												
Evaluate how the communication style(s) support or hinder the desired state of the process safety program												
With the Systems and Organizational team develop a plan to address gaps in the communication program												
Develop a plan with resource and budget implications for making changes to the communication style and systems												
Agree plan with Guiding Team and Exec Sponsor												
Implement plan												
P.S. Implementation Leads develop 'emerging issues' briefing or report												
Note - after having completed a period of time (perhaps 4-6 months) the P.S. Implementation Leads and Implementation Team Leads should take stock and develop what might be termed an interim or emerging issues briefing or report.												

TASKS	Months and Weeks 'Beginning'											
	Month 1				Month …				Month 12			
	1	2	3	4	…	…	…	…	50	51	52	53
Team leads in conjunction with Proj Director discuss and develop individual progress, issues, etc.												
These individual findings are collated into an interim or 'emerging issues' report												
P.S. Impl Lead and Team Leads agree the main topics, headings or 'emerging issues' as a Team												
Secretariat compiles this into a presentation and briefing paper.												
The P.S. Impl Lead and Team Leads present these to the Guiding Team												
Guiding Team to present to Board												
Consider whether to release to all staff and external stakeholders												
Transition Process												
Bining, sorting and prioritization of all observations/issues arising during implementation process												
Identify remaining gaps between current status and desired state												

TASKS	Months and Weeks 'Beginning'											
	Month 1				Month …				Month 12			
	1	2	3	4	…	…	…	…	50	51	52	53
Develop Report & Presentations for Guiding Team on remaining gaps												
Risk rank gaps												
Outline, possible continuing actions												
P.S. Impl Lead & Team Leads brief Guiding Team on findings and possible further work												
Guiding Team briefs Snr & Exec Management												
Agree responsibilities for carrying forward further integration activities												
Responsible line managers briefed												
Brief Stakeholder Liaison Group												

EXEMPLAR INTEGRATION BUDGET

		Hours	Hours	Hours	Hours	Hours	Cost Rate	
Name	**Unit/Org**	Week 1	Week 2	Week ...	Week 51	Week 52	**$/Hr**	**TOTAL HRS**
Exec Sponsor	Board/Exec level			8			1000	8
Guiding Team								
Snr. Impl. Mgr. 1	VP or Dir level			8			500	8
Snr. Impl. Mgr. 2	VP or Dir level		24	40			500	64
Snr. Impl. Mgr. 3	VP or Dir level		24	16			500	40
Snr. Impl. Mgr. 4	VP or Dir level		24	16			500	40
P.S. Impl Lead	Dir level		24	16			400	40
P.S. Impl Secretariat	Snr. Admin. Level		24	16			150	40
Sub-Total		0	120	120	0	0		240
Intolerable Risks & Physical Plant & Equipment Team								
Team Lead	Dir level		40	40			400	80
Engineering Rep	Mgr-Tech Spclst		40	40			300	80
Operations Rep	Mgr level						300	0
Maintenance Rep	Mgr level						300	0
Reliability Eng	Snr Tech Spclst						250	0

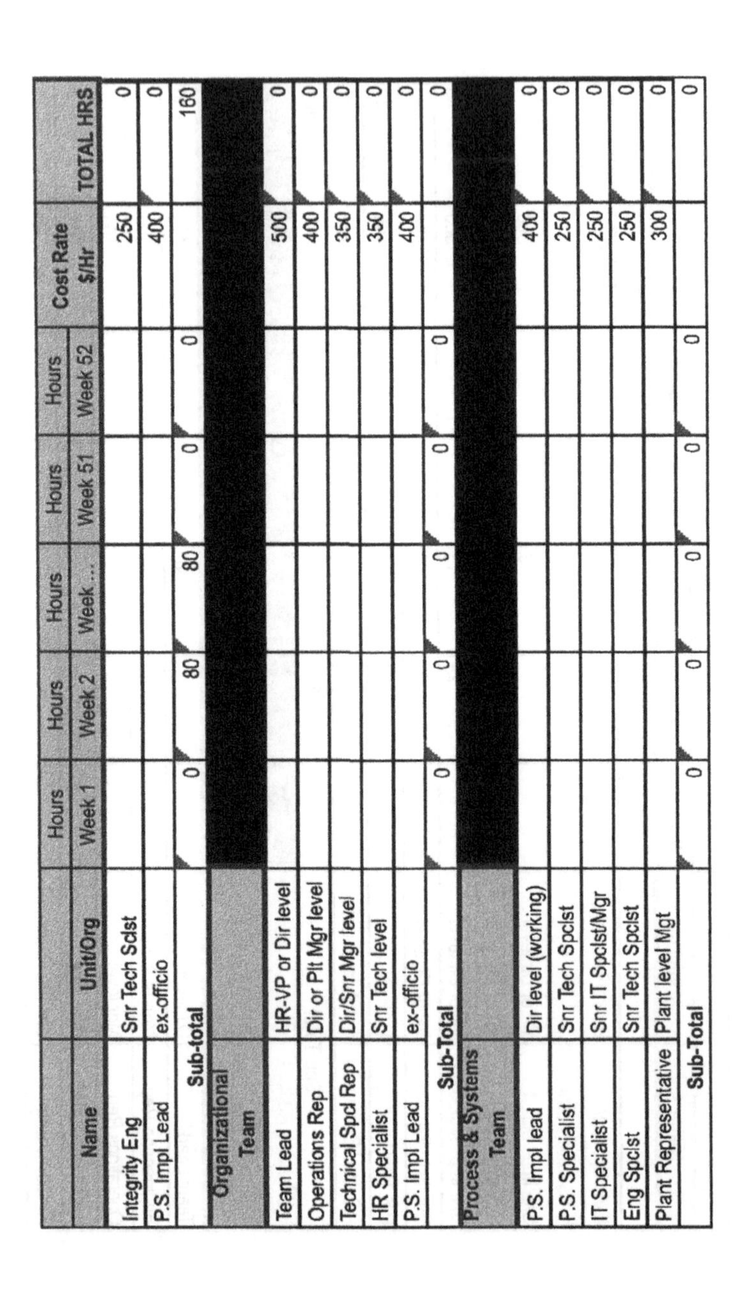

		Hours	Hours	Hours	Hours	Hours	Cost Rate	
Name	Unit/Org	Week 1	Week 2	Week ...	Week 51	Week 52	$/Hr	TOTAL HRS
Integrity Eng	Snr Tech Sclst						250	0
P.S. Impl Lead	ex-officio						400	0
Sub-total		0	80	80	0	0		160
Organizational Team								
Team Lead	HR-VP or Dir level						500	0
Operations Rep	Dir or Plt Mgr level						400	0
Technical Spcl Rep	Dir/Snr Mgr level						350	0
HR Specialist	Snr Tech level						350	0
P.S. Impl Lead	ex-officio						400	0
Sub-Total		0	0	0	0	0		0
Process & Systems Team								
P.S. Impl lead	Dir level (working)						400	0
P.S. Specialist	Snr Tech Spclst						250	0
IT Specialist	Snr IT Spclst/Mgr						250	0
Eng Spclst	Snr Tech Spclst						250	0
Plant Representative	Plant level Mgt						300	0
Sub-Total		0	0	0	0	0		0

		Hours	Hours	Hours	Hours	Hours	Cost Rate	
Name	Unit/Org	Week 1	Week 2	Week ...	Week 51	Week 52	$/Hr	TOTAL HRS
People Team								
Team Lead	Snr HR Mgr/Spclst						350	0
Technical Spcl Rep	Tech Mgr/Spclst						300	0
Plant Mgr Rep	Mgr or Sptd't level						300	0
P.S. Specialist	Snr P.S. Spclst						300	0
P.S. Impl Lead	ex-officio						400	0
Sub-Total		0	0	0	0	0		0
Contacts & Roll-0ut								
SPOC-Plant level	Plant level Mgt						300	0
SPOC-Plant level	Plant level Mgt						300	0
SPOC-Tech staff	Snr Tech Spclst						250	0
SPOC-IT systems	Snr IT Spclst/Mgr						300	0
Stkhldr Liasion	Snr Representative						300	0
Stkhldr Liasion	Snr Representative						300	0
Sub-Total		0	0	0	0	0		0
Total Hrs/Week		0	400	400	0	0		800
Cumulative Total Hours		0	400	800	800	800		

		Hours	Hours	Hours	Hours	Hours	Cost Rate	
Name	**Unit/Org**	Week 1	Week 2	Week …	Week 51	Week 52	**$/Hr**	**TOTAL HRS**
MANHOUR COSTS OR CHARGES								
Exec Sponsor	Board/Exec level	0	0	8000	0	0		8000
Guiding Team								
Snr. Impl. Mgr. 1	VP or Dir level	0	12000	20000	0	0		32000
Snr. Impl. Mgr. 2	VP or Dir level	0	12000	8000	0	0		20000
Snr. Impl. Mgr. 3	VP or Dir level	0	12000	8000	0	0		20000
Snr. Impl. Mgr. 4	VP or Dir level	0	9600	6400	0	0		16000
P.S. Impl Lead	Dir level	0	3600	2400	0	0		6000
P.S. Impl Secretariat	Snr. Admin. Level	0	16000	16000	0	0		32000
Sub-Total		0	65200	68800	0	0		134000
Intolerable Risks & Physical Plant & Equipment Team								
Team Lead	Dir level	0	0	0	0	0		0
Engineering Rep	Mgr-Tech Spclst	0	0	0	0	0		0
Operations Rep	Mgr level	0	0	0	0	0		0
Maintenance Rep	Mgr level	0	0	0	0	0		0
Reliability Eng	Snr Tech Spclst	0	0	0	0	0		0
Integrity Eng	Snr Tech Sclst	0	0	0	0	0		0
P.S. Impl Lead	ex-officio	0	0	0	0	0		0
Sub-Total		0	0	0	0	0		0

		Hours	Hours	Hours	Hours	Hours	Cost Rate	
Name	Unit/Org	Week 1	Week 2	Week ...	Week 51	Week 52	$/Hr	TOTAL HRS
Organizational Team								
Team Lead	HR-VP or Dir level	0	0	0	0	0		0
Operations Rep	Dir or Plt Mgr level	0	0	0	0	0		0
Technical Spcl Rep	Dir/Snr Mgr level	0	0	0	0	0		0
HR Specialist	Snr Tech level	0	0	0	0	0		0
P.S. Impl Lead	ex-officio	0	0	0	0	0		0
Sub-Total		0	0	0	0	0		0
Process & Systems Team								
P.S. Impl lead	Dir level (working)	0	0	0	0	0		0
P.S. Specialist	Snr Tech Spclst	0	0	0	0	0		0
IT Specialist	Snr IT Spclst/Mgr	0	0	0	0	0		0
Eng Spclst	Snr Tech Spclst	0	0	0	0	0		0
Plant Representative	Plant level Mgt	0	0	0	0	0		0
Sub-Total		0	0	0	0	0		0
People Team								
Team Lead	Snr HR Mgr/Spclst	0	0	0	0	0		0
Technical Spcl Rep	Tech Mgr/Spclst	0	0	0	0	0		0
Plant Mgr Rep	Mgr or Sptd't level	0	0	0	0	0		0
P.S. Specialist	Snr P.S. Spclst	0	0	0	0	0		0
P.S. Impl Lead	ex-officio	0	0	0	0	0		0
Sub-Total		0	0	0	0	0		0

		Hours	Hours	Hours	Hours	Hours	Cost Rate	
Name	Unit/Org	Week 1	Week 2	Week ...	Week 51	Week 52	$/Hr	TOTAL HRS
Contacts & Roll-0ut								
SPOC-Plant level	Plant level Mgt	0	0	0	0	0		0
SPOC-Plant level	Plant level Mgt	0	0	0	0	0		0
SPOC-Tech staff	Snr Tech Spclst	0	0	0	0	0		0
SPOC-IT systems	Snr IT Spclst/Mgr	0	0	0	0	0		0
Stkhldr Liasion	Snr Representative	0	0	0	0	0		0
Stkhldr Liasion	Snr Representative	0	0	0	0	0		0
Sub-Total		0	0	0	0	0		0
TOTAL MANHOUR COSTS/WEEK		0	130400	137600	0	0		268000
ASSOCIATED EXPENSES		Exp	Exp	Exp	Exp	Exp		Total-Exp
Exec Sponsor	Board/Exec level							0
Guiding Team								
Snr. Impl. Mgr. 1	VP or Dir level							0
Snr. Impl. Mgr. 2	VP or Dir level							0
Snr. Impl. Mgr. 3	VP or Dir level							0
Snr. Impl. Mgr. 4	VP or Dir level							0
P.S. Impl Lead	Dir level							0
P.S. Impl Secretariat	Snr. Admin. Level							0
Sub-Total								0

		Hours	Hours	Hours	Hours	Hours	Cost Rate	
Name	**Unit/Org**	Week 1	Week 2	Week ...	Week 51	Week 52	**$/Hr**	**TOTAL HRS**
Intolerable Risks & Physical Plant & Equipment Team								
Team Lead	Dir level							0
Engineering Rep	Mgr-Tech Spclst							0
Operations Rep	Mgr level							0
Maintenance Rep	Mgr level							0
Reliability Eng	Snr Tech Spclst							0
Integrity Eng	Snr Tech Sclst							0
P.S. Impl Lead	ex-officio							0
Sub-Total								0
Organizational Team								
Team Lead	HR-VP or Dir level							0
Operations Rep	Dir or Plt Mgr level							0
Technical Spcl Rep	Dir/Snr Mgr level							0
HR Specialist	Snr Tech level							0
P.S. Impl Lead	ex-officio							0
Sub-Total								0
Process & Systems Team								
P.S. Impl lead	Dir level (working)							0
P.S. Specialist	Snr Tech Spclst							0

Name	Unit/Org	Hours Week 1	Hours Week 2	Hours Week …	Hours Week 51	Hours Week 52	Cost Rate $/Hr	TOTAL HRS
IT Specialist	Snr IT Spclst/Mgr							0
Eng Spclst	Snr Tech Spclst							0
Plant Representative	Plant level Mgt							0
Sub-Total								0
People Team								
Team Lead	Snr HR Mgr/Spclst							0
Technical Spcl Rep	Tech Mgr/Spclst							0
Plant Mgr Rep	Mgr or Sptd't level							0
P.S. Specialist	Snr P.S. Spclst							0
P.S. Impl Lead	ex-officio							0
Sub-Total								0
Contacts & Roll-0ut								
SPOC-Plant level	Plant level Mgt							0
SPOC-Plant level	Plant level Mgt							0
SPOC-Tech staff	Snr Tech Spclst							0
SPOC-IT systems	Snr IT Spclst/Mgr							0
Stkhldr Liasion	Snr Representative							0
Stkhldr Liasion	Snr Representative							0
Sub-Total								0
TOTAL EXPENSES		0.00	0.00	0.00	0.00	0.00		0
Grand Total Fees & Expenses		0.00	130400.00	137600.00	0.00	0.00		**268000**

REFERENCES

[1] H.H. Fawcett and W.S. Wood, Safety and Accident Prevention in Chemical Operations, 2nd ed. John Wiley and sons, New York: 1982 p.1.
[2] Center for Chemical Process Safety (CCPS), Process Safety Leading and Lagging Metrics, Initial Release, American Institute of Chemical Engineers, New York, New York, December 20, 2007.
[3] Center for Chemical Process Safety (CCPS), Guidelines for Risk Based Process Safety, American Institute of Chemical Engineers, New York, New York, 2007.
[4] American Petroleum Institute, 1220 L Street, NW, Washington, DC 20005. www.api.org
[5] American Chemistry Council, 1300 Wilson Blvd., Arlington, VA 22209. www.americanchemistry.com
[6] International Organization for Standardization (ISO); ISO 14001 – Environmental Management System, Geneva, Switzerland. www.iso.org/iso/en/iso9000-14000/index.html
[7] International Organization for Standardization (ISO); OHSAS 18001 – International Occupational Health and Safety Management System, Geneva, Switzerland; www.ohsas-18001-occupational-health-and-safety.com/
[8] Organization for Economic Cooperation and Development – Guiding Principles on Chemical Accident Prevention, Preparedness, and Response, 2nd edition, 2003, Organisation for Economic Co-Operation and Development, Paris, 2003. www2.oecd.org/guidingprinciples/index.asp
[9] American National Standards Institute, 25 West 43rd Street, New York, NY 10036. www.ansi.org
[10] American Society of Mechanical Engineers, Three Park Avenue, New York, NY 10016. www.asme.org
[11] The Chlorine Institute, 1300 Wilson Blvd., Arlington, VA 22209. www.chlorineinstitute.org
[12] The Instrumentation, Systems, and Automation Society/International Electrotechnical Commission, 67 Alexander Drive, Research Triangle Park, NC 27709. www.isa.org
[13] National Fire Protection Association, 1 Batterymarch Park, Quincy, MA 02169. www.nfpa.org
[14] U.S. Occupational Safety and Health Administration; Process Safety Management of Highly Hazardous Chemicals Regulations, (29 CFR 1910.119), May 1992. www.osha.gov
[15] Section 5(a)(1) – General Duty Clause, Occupational Safety and Health Act of 1970, Public Law 91-596, 29 USC 654, December 29, 1970. www.osha.gov
[16] U.S. Environmental Protection Agency; Accidental Release Prevention Requirements: Risk Management Programs Under Clean Air Act Section 112(r)(7), 40 CFR 68, June 20, 1996 Fed. Reg. Vol. 61[31667-31730]. www.epa/gov
[17] Clean Air Act Section 112(r)(1) – Prevention of Accidental Releases – Purpose and general duty, Public Law No. 101-549, November 1990. www.epa.gov
[18] California Office of Emergency Services, California Accidental Release Program (CalARP) Regulation, CCR Title 19, Division 2 –Chapter 4.5, June 28, 2004. www.oes.ca.gov

---

[19] New Jersey Department of Environmental Protection, Toxic Catastrophe Prevention Act (TCPA), Bureau of Chemical Release Information and Prevention, N.J.A.C. 7:31 Consolidated Rule Document, April 17, 2006. www.nj.gov/dep

[20] Contra Costa County Industrial Safety Ordinance. www.co.contra-costa.ca.us/

[21] Delaware Department of Natural Resources and Environmental Control; Extremely Hazardous Substances Risk Management Act, Regulation 1201 Accidental Release Prevention Regulation, , March 11, 2006. www.dnrec.delaware.gov/

[22] Nevada Division of Environmental Protection, Chemical Accident Prevention Program (CAPP), NRS 459.380, February 15, 2005. http://ndep.nv.gov/bapc/capp/

[23] Australian National Occupational Health and Safety Commission; Australian National Standard for the Control of Major Hazard Facilities, NOHSC: 1014, 2002. www.docep.wa.gov.au/

[24] Environment Canada, Environmental Emergency Regulations (SOR/2003-307), www.ec.gc.ca/CEPARegisty/regulations/detailReg.cfm?intReg=70

[25] Control of Major-Accident Hazards Involving Dangerous Substances, European Directive Seveso II (96/82/EC). http://europa.eu.int/comm/environment/seveso

[26] Korean Ministry of Environment; Korean OSHA PSM standard, Industrial Safety and Health Act – Article 20, Preparation of Safety and Health Management Regulations. – Framework Plan on Hazardous Chemicals Management, 2001 – 2005. www.kosha.net/isp/board/viewlist.jsp?cf=29099&x=19565&no=3

[27] Malaysia – Department of Occupational Safety and Health (DOSH) Ministry of Human Resources Malaysia, Section 16 of Act 514. http://dosh.mohr.gov.my/

[28] Mexican Integral Security and Environmental Management System (SIASPA), 1998. www.pepsonline.org/Publications/pemex.pdf

[29] UK- Health & Safety Executive; Control of Major Accident Hazards Regulations (COMAH), 1999 and 2005. www.hse.gov.uk/comah/

[30] United States Chemical Safety and Hazard Investigation, "Imperial Sugar Company Explosion and Fire" Port Wentworth, GA, February 7, 2008 , http://www.chemsafety.gov

[31] United States Chemical Safety and Hazard Investigation Board, "Combustible Dust Hazard Study", REPORT NO. 2006-H-1, November 2006

[32] American Petroleum Institute; API Recommended Practice 753, Management of Hazards Associated with Location of Process Plant Portable Buildings, First Edition, 1220 L Street NW, Washington, DC, June 2007.

[33] United States Chemical Safety and Hazard Investigation Board (USCSB), A Blast Wave in Danvers, August 25, 2008, video posted on CSB website: www.chemsafety.gov.

[34] Kletz, Trevor, *Still Going Wrong,* Elsevier Publishing Co. 2003

[35] Crowl, Daniel A., Louvar, Joseph F., Chemical Process Safety Fundamentals with Applications, Prentice Hall, 2008.

[36] Young & Partners, cited in Chemical Week, March 6, 2002, p. 28

[37] Kamakura, Yasuhiko "Corporate structural change and social dialogue in the chemical industry", International Labour Office, CH-1211 Geneva 22, Switzerland

[38] U.S. Chemical Safety and Hazards Investigation Board, Investigation Report – Refinery Explosion and Fire, BP Texas City, Report No. 2005-04-I-TX, March 2007

[39] Center for Chemical Process Safety (CCPS), Process Safety Leading and Lagging Metrics, Initial Release, American Institute of Chemical Engineers, New York, New York, December 20, 2007.

---

[40] Center for Chemical Process Safety, "Guidelines for Auditing Process Safety Management Systems, Second Edition", American Institute of Chemical Engineers, New York, New York, scheduled for publication in 2009.
[41] Center for Chemical Process Safety (CCPS), Process Safety Leading and Lagging Metrics, Initial Release, American Institute of Chemical Engineers, New York, New York, December 20, 2007.
[42]Kotter, J.; *Leading Change*, Harvard Business School Press, January 1996.
[43] Kotter, J & Cohen, D.; *The Heart of Change,* Perseus Distribution Svcs, January 2002.
[44] Pascale, R. & Athos, A.; *The Art of Japanese Management for American Executives*, Grand Central Publishing, January 1982,
[45] Koontz, H. & O'Donnell, C: *Principles of Management: An Analysis of Managerial Functions,* McGraw Hill Publishing, 1972
[46] Bremmer, I and Pujadas, J; "State Capitalism Makes a Comeback", Harvard Business Review, February 2009.
[47] Kumar, N.; "How Emerging Giants are Rewriting the Rules of M&A", Harvard Business Review, May 2009.
[48] Center for Chemical Process Safety (CCPS), Process Safety Leading and Lagging Metrics, Initial Release, American Institute of Chemical Engineers, New York, New York, December 20, 2007

INDEX

Printed and bound by CPI Group (UK) Ltd, Croydon, CR0 4YY

23/04/2025

14660904-0005